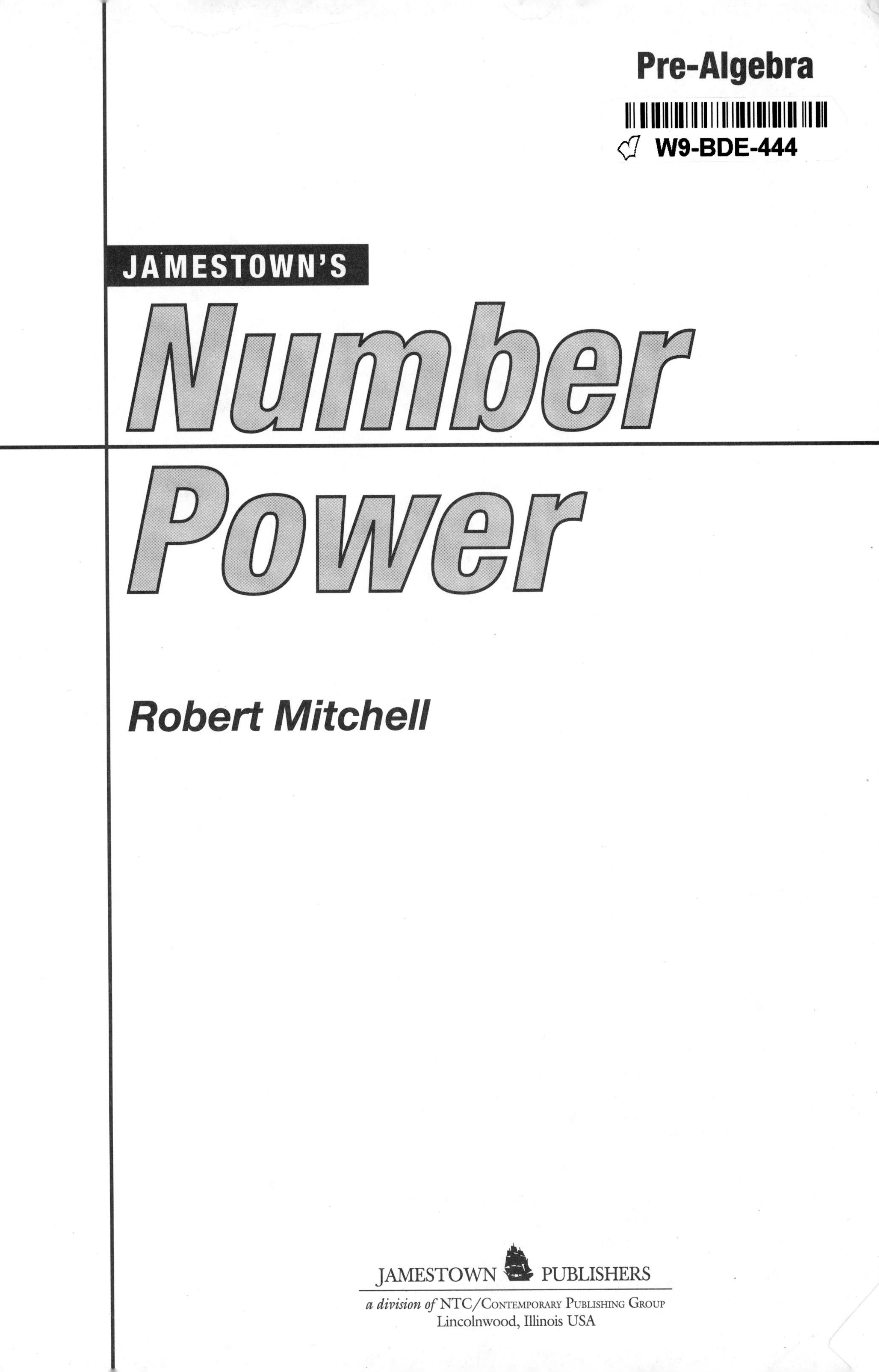

Pre-Algebra

JAMESTOWN'S

Number Power

Robert Mitchell

JAMESTOWN PUBLISHERS
a division of NTC/Contemporary Publishing Group
Lincolnwood, Illinois USA

ISBN: 0-8092-2283-3

Published by Jamestown Publishers,
a division of NTC/Contemporary Publishing Group, Inc.,
4255 West Touhy Avenue,
Lincolnwood (Chicago), Illinois 60712-1975 U.S.A.

Manufactured in the United States of America.

1 2 3 4 5 6 7 8 9 VH 16 15 14 13 12 11 10 9 8 7 6 5 4 3 2 1

Table of Contents

Problem-Solving Strategies

Data Analysis

Probability

Geometry and Measurement

Spatial Sense and Patterns

The Language of Algebra

To the Student

Welcome to *Pre-Algebra.*

Pre-Algebra is designed to strengthen your problem-solving and reasoning skills in basic mathematics. Important topics include number skills, word problem and reasoning skills, data analysis, probability, geometry, measurement, spatial sense, patterns, and beginning algebra. An understanding of these topics is necessary in many occupations. Also, these topics are included on educational and vocational tests, including GED, college entrance, civil service, and military enlistment tests.

This workbook is designed to prepare you for taking a test and for pursuing further education or training in your chosen field. The first part of the book, Building Number Power, is divided into eight chapters. Each chapter provides step-by-step instruction and gives you practice with your new skills. Each chapter ends with a chapter review to check your understanding.

The second part of the book, Using Number Power, gives you more practice applying pre-algebra skills. These applications are fun and are examples of the use of mathematics in everyday life.

Skills that you will find helpful in your study of pre-algebra are estimating and using a calculator. A summary page on estimation and mental math is provided on page 199. A summary page of calculator use is provided on page 198. Also, for quick reference, a summary of useful formulas is provided on pages 200 and 201. The following icons will alert you to problems where calculator, estimation, or mental math use may be especially helpful.

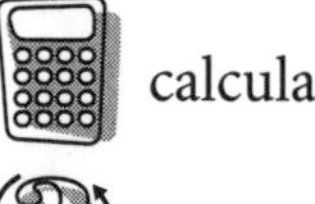 calculator icon

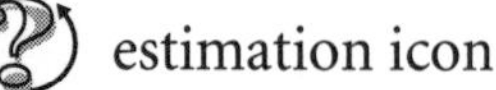 estimation icon

mental math icon

To get the most out of your work, do each problem carefully. A chart inside the back cover will help you keep track of your scores.

Pre-Algebra Pretest

This pretest will tell you which sections of *Pre-Algebra* you need to work on and which you have already mastered. Do every problem that you can. There is no time limit. Then use the chart at the end of the test to find the pages where you need work.

1. On a number line, what number is 7 units to the right of –4?

2. Circle each number that is a possible value of x in the inequality $x \leq 6$.

–3 6 0 9 3

3. Graph the following inequality on the number line.

$n \leq 3$

–6 –5 –4 –3 –2 –1 0 1 2 3 4 5 6

4. Find the value of the expression below.

$4(6 - 1)$

5. On her math test, Basha got 30 questions correct out of 39. *Estimate* the percent of test questions Basha got correct.

6. Greg wants to decrease his daily calories by 30%. He now eats 3,000 calories each day. Which expression tells the number of calories Greg is allowed on his diet?

a. $0.3 \times 3{,}000$
b. $3{,}000 - 30 \times 3{,}000$
c. $3{,}000 + 0.3 \times 3{,}000$
d. $3{,}000 - 0.3 \times 3{,}000$

7. In Wendi's aerobics class, there are 30 students. If 22 of the students are women, what is the ratio of women students to men students?

8. Community Hospital employs 85 nurses, 2 out of 5 of whom are men. Write a proportion that can be used to find the number of nurses (n) who are men.

9. From his home on Fir Street, Landon walked 1.6 miles east. He then turned left on 4th Avenue and walked 0.8 miles north. Next, he turned left and walked 2.3 miles west. How far is he from home if he now turns left and walks to Fir Street and then walks down Fir Street to his home?

A Drawing May Help

10. Kaitlin is going to have a sandwich, drink, and dessert for lunch. For a sandwich, she can choose cheese or chicken. For a drink, she can choose milk or soda. For dessert, she can choose ice cream or cake. How many different lunch combinations are available to Kaitlin?

A List May Help

Sandwich **Drink** **Dessert**

11. Brent bought five pizzas for the class party. The bill came to $54.00. He bought two sizes of pizzas: small pizzas for $9.50 each and large pizzas for $12.75 each. How many of each size pizza did Brent buy?

Small pizzas: ____________

Large pizzas: ____________

A Table May Help

Small Pizza $9.50	Large Pizza $12.75	Total Cost

Guessing and Checking May Help

12. Five members of the Emerson family paid a total of $14.00 for admission to the Lynn Valley Carnival. How many of each type of ticket did the Emersons buy?

Adult: ______

Student: ______

Child: ______

Lynn Valley Carnival	
Admission:	
Adult	$4.50
Student	$3.00
Child (under 6)	$1.75

In a survey, downtown shoppers were asked, "How many times each week do you shop downtown?" For problems 13 and 14, refer to the responses listed on the tally sheet.

13. What is the ratio of the number of shoppers who shop 4 or more times each week to those who shop 1 time?

Survey of Downtown Shoppers

1 time	2 times	3 times	4 or more times
𝍸 𝍸 ||||	𝍸 𝍸 𝍸 ||||	𝍸 𝍸 𝍸 𝍸 𝍸 𝍸	𝍸 ||

14. *About* what fraction of all shoppers shop downtown three times each week?

For problems 15 and 16, refer to the bar graph at the right. Each student in Annette's class voted for one favorite pet.

15. How many more students chose a dog as their favorite pet than a bird?

16. What percent of students chose a hamster as their favorite pet?

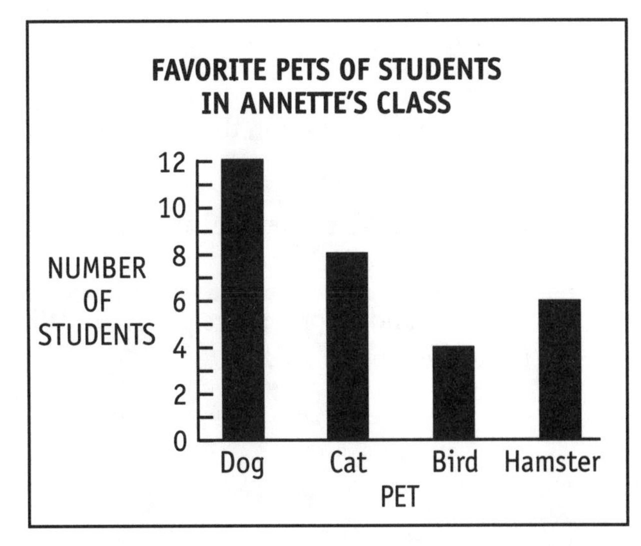

17. A coin purse contains three pennies, two nickels, and four dimes. Suppose you reach into the purse and randomly take out one coin. What is the probability that the coin you choose will be a nickel?

For problems 18 and 19, refer to the spinner shown at the right.

18. If you spin the pointer once, what is the probability that the pointer will stop on a section labeled $25?

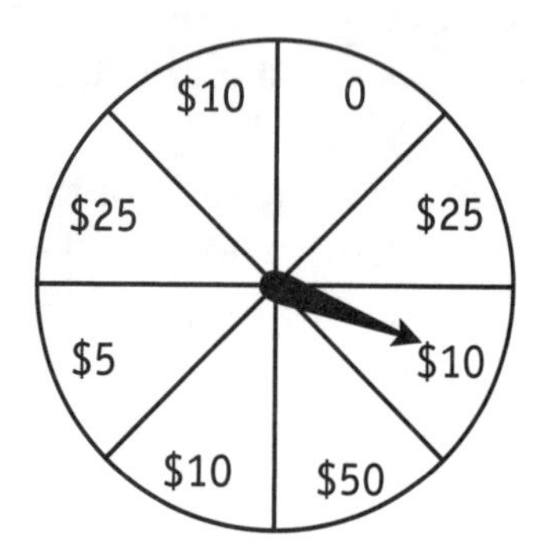

19. Suppose you spin the pointer 60 times. About how many times is the pointer likely to stop on a section labeled $25?

20. Suppose, with your eyes closed, you choose a penny from the group shown at the right. You then choose a second penny. What is the probability that both pennies you choose will be heads up?

21. In the drawing at the right, the measure of $\angle x$ is 42°.

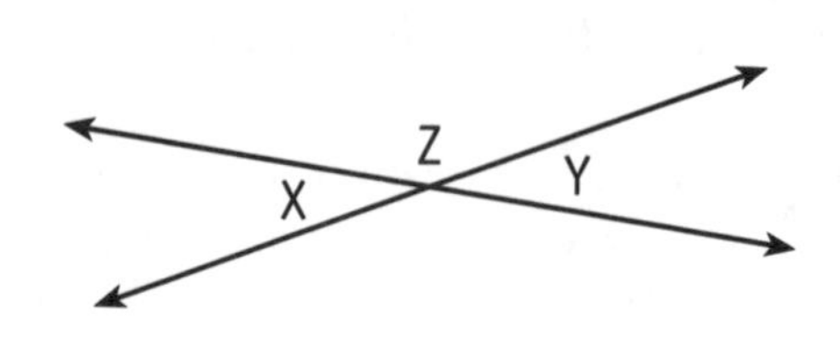

a. Angles X and Y are vertical angles. What is the measure of $\angle Y$?

b. Angles Z and X form a straight angle. What is the measure of $\angle Z$?

22. In ΔLMN, the measures of ∠M and ∠N are given. What is the measure of ∠L?

L

52°

M

95°

N

23. A radio antenna broadcasts its signal in a circular pattern. The maximum distance the station can send its signal is 80 miles. What is the approximate area over which the station can be heard? (Use 3.14 for π.)

Maximum range = 80 miles

24. A swimming pool is 62.5 feet long, 32.5 feet wide, and has an average depth of 5.5 feet.

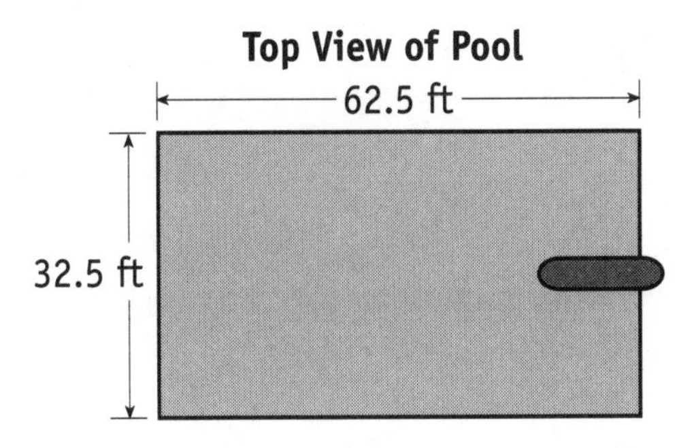

a. To the nearest 0.1 (tenth) cubic foot, how much water does the pool hold when full?

b. To the nearest 1,000 gallons, how many gallons of water does the pool hold when full? (1 cubic foot ≈ 7.5 gallons)

25. Draw each line of symmetry for the rectangle shown at the right.

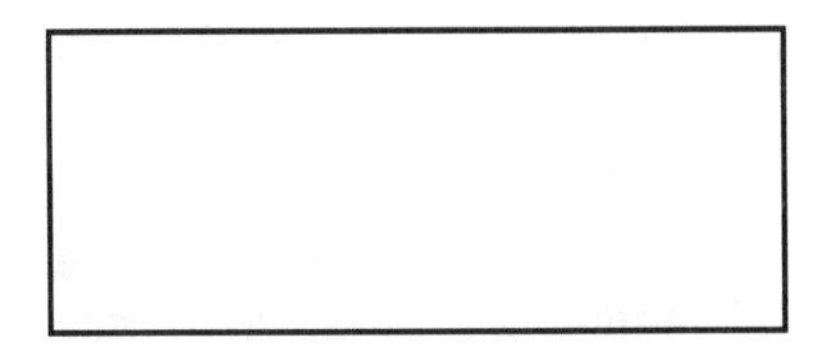

26. **a.** Draw two rectangles that are congruent.

b. Draw two rectangles that are similar but not congruent.

For problem 27, refer to the grid at the right.

27. Points A, B, C, and D are the corners (vertices) of a rectangle. Write the coordinates of each of these points.

A = (,)

B = (,)

C = (,)

D = (,)

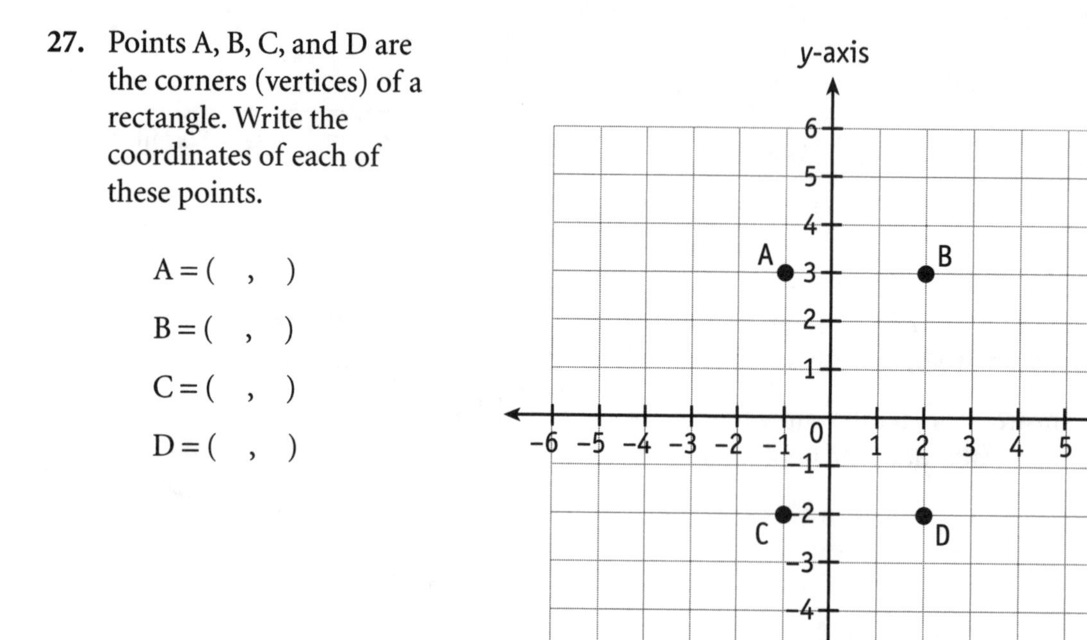

28. How many short line segments does it take to draw the fourth figure in the pattern shown below?

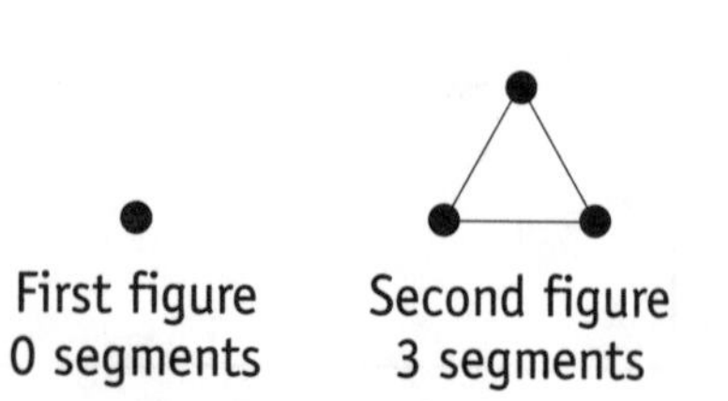

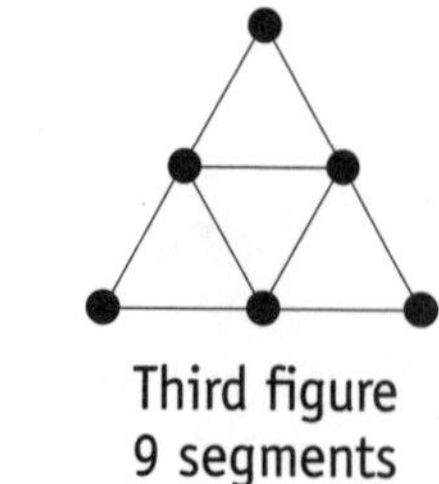

29. The newspaper says that rent has increased an average of 8% this past year. Write an algebraic expression that represents this year's average rent if last year's average rent was $650 per month.

30. What is the value of the algebraic expression below when $x = 3$, $y = 2$, and $z = 5$?

$$3x^2 + y^2 - 6z$$

31. Solve the subtraction equation below. Show each step of your solution.

$$x - 7 = 16$$

32. Solve the multiplication equation below. Show each step of your solution.

$$3y = 27$$

Pre-Algebra Pretest Chart

If you miss more than one problem in any section of this test, you should complete the lessons on the practice pages indicated on the chart below. If you get all the problems correct in a section of the test, you probably do not need further study in that chapter. However, we recommend that you read through that chapter to be sure you understand all the pre-algebra skills taught there.

PROBLEM NUMBERS	SKILL AREA	PRACTICE PAGES
1	positive and negative numbers	16–17
2, 3	inequalities	18–21
4	order of operations	22–23
5	estimating	32–37
6	set-up questions	38–41
7	ratio	44–47
8	proportion	48–51
9	drawing a picture	54–55
10	making a list	56–57
11	making a table	58–59
12	guess and check	60–61
13, 14	using a tally sheet	71
15, 16	reading a bar graph	78–79
17, 18, 19	probability	90–95
20	probability of two events	102–103
21	angles	107–111
22	triangles	112–113
23	area of a circle	118
24	volume of a rectangular solid	120
25	symmetry	124–125
26	congruent and similar figures	126
27	coordinate grid	130–133
28	geometric patterns	136–137
29	writing an algebraic expression	142–143
30	evaluating algebraic expression	146–147
31, 32	solving equations	150–157

Building Number Power

Reviewing Fractions, Decimals, and Percents

Numbers Less than 1

A number less than 1 represents part of a whole. For example, the circle at the right is divided into 10 equal parts, 7 of which are shaded. The shaded part of the circle can be described in three ways.

- as a fraction: $\frac{7}{10}$ is shaded
- as a decimal: 0.7 is shaded
- as a percent: 70% is shaded

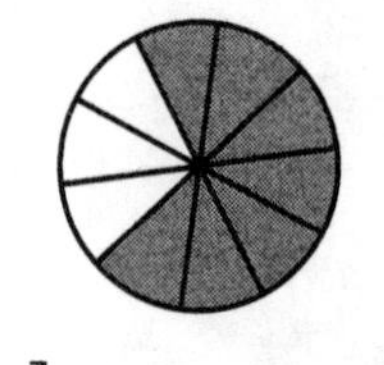

$\frac{7}{10} = 0.7 = 70\%$

Reviewing Your Skills

On the next six pages, you will check your basic skills with fractions, decimals, and percents. For a thorough review of these skills, see *Number Power Fractions, Decimals, and Percents.*

Reviewing Fractions

Write the proper fraction and word name for each shaded part.

	Whole Object	Description	Fraction	Word Name
1.		1 of 2 equal parts	$\frac{1}{2}$	*one half*
2.		1 of 3 equal parts	______	______
3.		1 of 4 equal parts	______	______
4.		1 of 5 equal parts	______	______
5.		1 of 6 equal parts	______	______
6.		1 of 8 equal parts	______	______
7.		1 of 10 equal parts	______	______
8.		1 of 100 equal parts	______	______

Reviewing Decimals

Write each decimal in words.

1. 0.2 *two tenths*

2. 0.6 ______

3. 0.9 ______

4. 0.05 ______

5. 0.30 ______

6. 0.45 ______

7. 0.008 ______

8. 0.035 ______

9. 0.150 ______

10. 0.325 ______

Write each part as a decimal.

11. three tenths *0.3*

12. seven tenths ______

13. five hundredths ______

14. forty hundredths ______

15. seventy-three hundredths ______

16. fifty-six hundredths ______

17. five thousands ______

18. eighty-two thousandths ______

19. one hundred four thousandths ______

20. two hundred sixty-seven thousandths ______

Write each fraction as a decimal.

21. $\frac{7}{10}$ $\frac{1}{2}$ $\frac{2}{5}$ $\frac{3}{8}$ $\frac{3}{4}$

Write each decimal as a fraction in lowest terms.

22. 0.25 0.42 0.64 0.50 0.90

Reviewing Percent

Remember:
Percent means *parts out of 100.*

In each problem, write the

- **percent of each grid that is shaded**
- **percent of each grid that is not shaded**
- **sum (total) of the two percents**

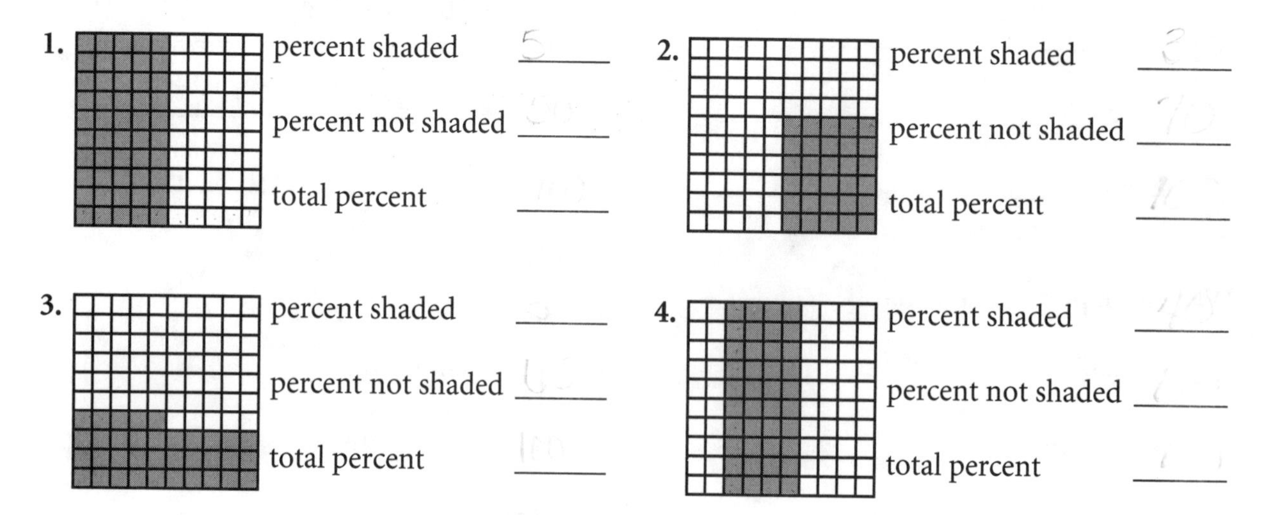

What percent of $1.00 is each group of coins below?

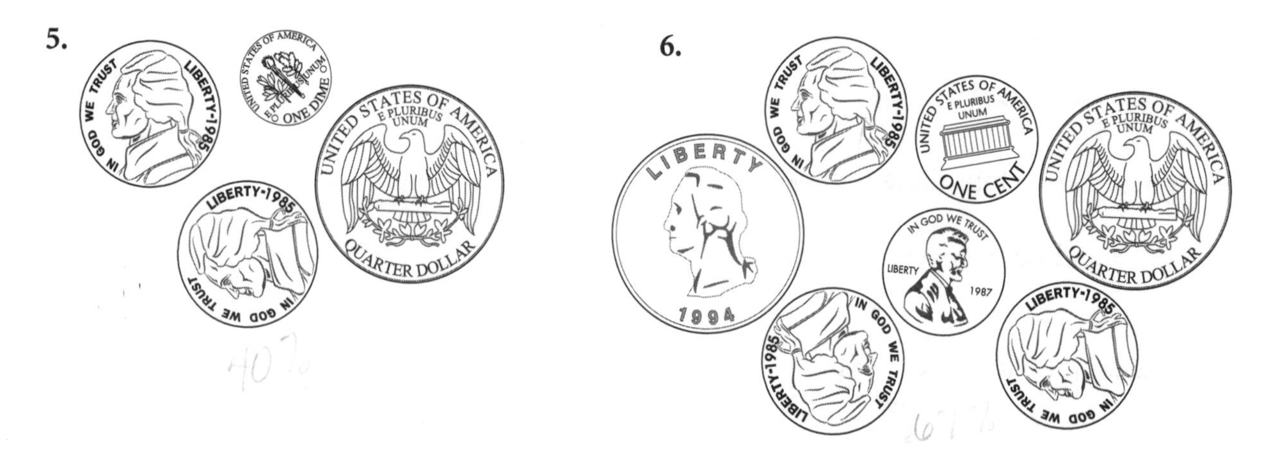

7. Shade 30% of the grid.

8. Shade 19% of the grid.

Relating Fractions, Decimals, and Percents

Fractions, decimals, and percents can be changed into one another.

- A fraction can be changed to a percent by multiplying by 100%. EXAMPLE $\frac{1}{4} \times 100\% = 25\%$
- A decimal can be changed to a percent by moving the decimal point two places to the right. EXAMPLE $0.37 = 37\%$
- Percent has the same value as a fraction with a denominator of 100. EXAMPLE $37\% = \frac{37}{100}$
- Percent has the same value as a two-place decimal. EXAMPLE $37\% = 0.37$

Did You Know . . . ?

$\frac{1}{3} = 33\frac{1}{3}\%$

$\frac{2}{3} = 66\frac{2}{3}\%$

You'll want to remember these.

Write each fraction as a percent.

1. $\frac{3}{4}$ $\frac{1}{2}$ $\frac{7}{10}$ $\frac{4}{5}$ $\frac{7}{20}$

Write each decimal as a percent.

2. 0.13 0.25 0.37 0.82 0.5

Write each percent as a fraction in lowest terms.

3. 17% 35% 40% 75% 82%

Write each percent as a decimal.

4. 30% 50% 42% 60% 95%

Answer each question.

5. One meter is equal to 100 centimeters.

a. What fraction of 1 meter is 35 centimeters? ______

b. What percent of 1 meter is 58 centimeters? ______

6. A sale advertises 30% off all sweaters.

a. Write the discount percent as a fraction. ______

b. Write the discount percent as a decimal. ______

Reviewing Computation Skills

FRACTIONS

Add or subtract.

1. $\frac{1}{3} + \frac{1}{3}$ $\qquad$ $\frac{3}{5} + \frac{2}{5}$ $\qquad$ $\frac{7}{8} + \frac{5}{8}$ $\qquad$ $\frac{1}{2} + \frac{1}{4}$ $\qquad$ $\frac{3}{4} + \frac{5}{8}$

2. $\frac{7}{8} - \frac{2}{8}$ $\qquad$ $\frac{3}{4} - \frac{1}{4}$ $\qquad$ $\frac{5}{6} - \frac{1}{2}$ $\qquad$ $1\frac{1}{4} - \frac{3}{4}$ $\qquad$ $1\frac{3}{8} - \frac{1}{2}$

Multiply.

3. $\frac{1}{2} \times \frac{1}{2} =$ $\qquad$ $\frac{1}{3} \times \frac{3}{5} =$ $\qquad$ $\frac{3}{4} \times \frac{2}{3} =$ $\qquad$ $\frac{7}{8} \times \frac{3}{2} =$

4. $\frac{1}{2} \times 5 =$ $\qquad$ $6 \times \frac{2}{3} =$ $\qquad$ $1\frac{1}{2} \times 3 =$ $\qquad$ $3 \times 2\frac{1}{3} =$

Divide.

5. $\frac{1}{2} \div \frac{1}{4} =$ $\qquad$ $\frac{1}{3} \div \frac{1}{2} =$ $\qquad$ $\frac{3}{4} \div \frac{2}{3} =$ $\qquad$ $\frac{1}{8} \div \frac{1}{8} =$

6. $\frac{1}{4} \div 2 =$ $\qquad$ $\frac{1}{2} \div 3 =$ $\qquad$ $4 \div \frac{1}{2} =$ $\qquad$ $6 \div \frac{2}{3} =$

DECIMALS

Add or subtract.

7. $\begin{array}{r} 6.4 \\ +\,3.2 \\ \hline \end{array}$ $\begin{array}{r} 10.8 \\ +\;\;2.95 \\ \hline \end{array}$ $\begin{array}{r} \$14.06 \\ +\;\;\;5.90 \\ \hline \end{array}$ $3.75 + 1.5 =$ $\$10.60 + \$8.25 =$

8. $\begin{array}{r} 7.3 \\ -\,5.1 \\ \hline \end{array}$ $\begin{array}{r} 9.3 \\ -\,4.7 \\ \hline \end{array}$ $\begin{array}{r} \$5.72 \\ -\;\;1.05 \\ \hline \end{array}$ $7.46 - 1.6 =$ $\$12.70 - \$5.85 =$

Multiply or divide.

9. $\begin{array}{r} 2.4 \\ \times\,5 \\ \hline \end{array}$ $\begin{array}{r} 3.5 \\ \times\,1.4 \\ \hline \end{array}$ $\begin{array}{r} \$4.76 \\ \times\,0.03 \\ \hline \end{array}$ $\begin{array}{r} 2.25 \\ \times\,8.9 \\ \hline \end{array}$ $\begin{array}{r} \$6.75 \\ \times\;2.5 \\ \hline \end{array}$

10. $4\overline{)1.6}$ $3\overline{)0.06}$ $1.2\overline{)2.88}$ $2.6\overline{)\$0.52}$ $0.15\overline{)\$7.50}$

PERCENTS

Find each part.

11. 25% of 40 = 10 14% of 34 4.76 5.5% of \$200 11.00

Answer each question.

12. What percent of 25 is 15?

13. What percent of 300 is 12?

14. 20% of what number is 13?

15. \$24 is 10% of what number?

Positive and Negative Numbers

Our number system uses both positive and negative numbers.

- **Positive numbers** are greater than zero.
- **Negative numbers** are less than zero.
- **Zero** itself is neither positive nor negative.

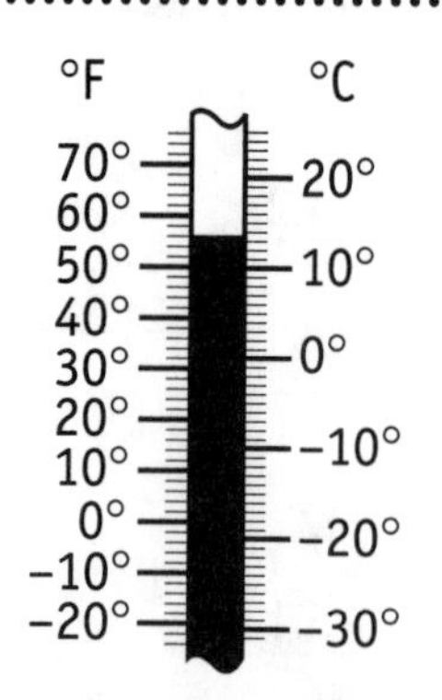

A thermometer shows both positive and negative temperatures.

A thermometer shows both positive and negative temperatures. As seen on the thermometer, negative values are always preceded by a negative (minus) sign "–".

- 20° below 0° is written –20°.
- A $3\frac{1}{2}$ dollar drop in a stock market price is written $-3\frac{1}{2}$.
- 85 feet below sea level is written as an altitude of –85 feet.

A **number line** shows both positive and negative numbers. On a **horizontal number line,** negative numbers are to the left of zero; positive numbers are to the right.

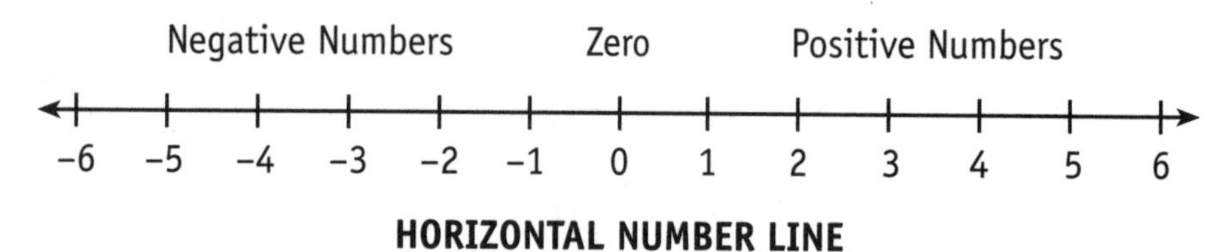

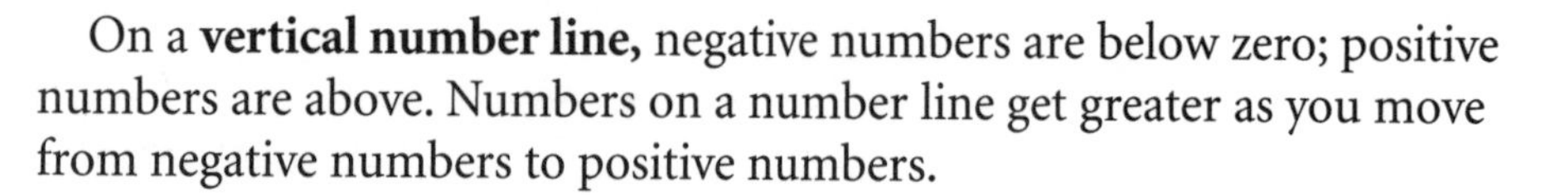

HORIZONTAL NUMBER LINE

On a **vertical number line,** negative numbers are below zero; positive numbers are above. Numbers on a number line get greater as you move from negative numbers to positive numbers.

- Every negative number is less than 0.
- Every positive number is greater than 0.
- Every positive number is greater than every negative number.

A negative number can be read in any of three ways.
You can read –5 as

- negative 5
- minus 5
- 5 below zero

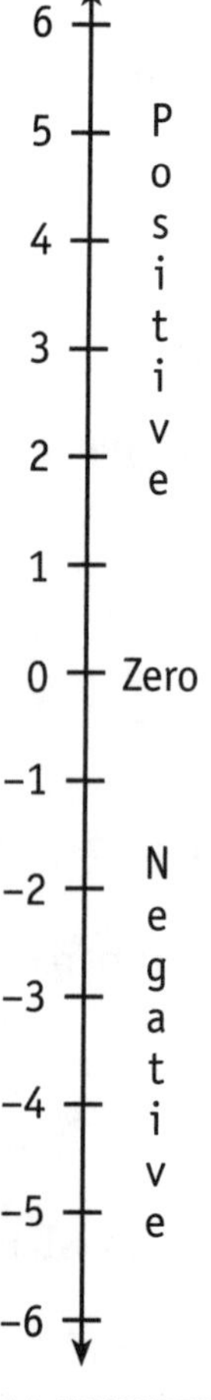

VERTICAL NUMBER LINE

1. Write letters above the number line to show where each number belongs. Point A is done as an example.

A 3 **B** –4 **C** $1\frac{1}{2}$ **D** 5.25 **E** $-3\frac{1}{2}$ **F** 0.75 **G** –5.55

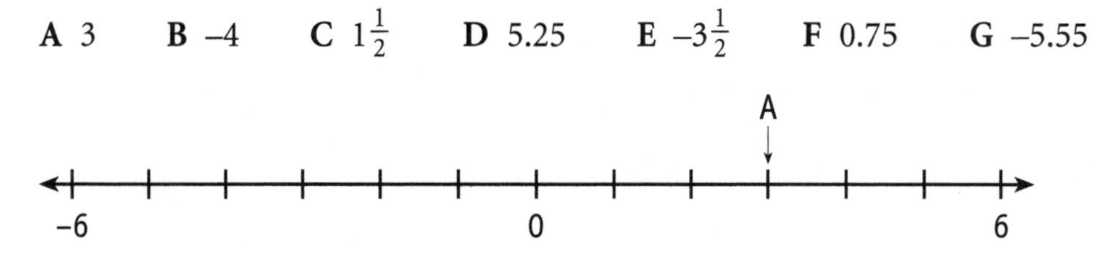

2. **a.** Write the letter H above the point –2 on the number line below.

–6 0 6

b. Write the letter I above the point that is 4 units to the right of point H.

c. Write the letter J above the point that is 3 units to the right of point I.

d. Write the letter K above the point that is 9 units to the left of point J.

e. What number does each letter represent?

I is ______ J is ______ K is _______

3. Circle the greatest number in each group.

a. –3, 0, –2 4, 0, –2 5, 3, –7 –4, –2, –8

b. –1, –5, –6 3, –1, –5 –6, –2, 0 –5, 3, –1

4. Circle the least number in each group.

a. –8, –10, 0 7, 0, 2 8, –1, 0 –2, 6, –4

b. –4, 7, –5 0, 3, –2 –2, 4, 0 0, 1, –1

Comparing Numbers: Inequalities

An **inequality** is a comparison of two or more numbers.
Four comparison symbols are used with inequalities.

Symbol	Meaning	Example	Read as
$<$	is less than	$n < 8$	*n* is less than 8
$>$	is greater than	$m > 10$	*m* is greater than 10
$\leq$	is less than *or* equal to	$r \leq 12$	*r* is less than *or* equal to 12
$\geq$	is greater than *or* equal to	$s \geq 15$	*s* is greater than *or* equal to 15
$\neq$	is not equal to	$p \neq 7$	*p* is not equal to 7

In each inequality, the letter can be *any value* that makes the inequality true.

- $n < 8$ *n* can be any value less than 8.
  Example values: $n = 5$, $n = 0$ or $n = -7.5$.
- $m > 10$ *m* can be any value greater than 10.
  Example values: $m = 10.5$, $m = 11$, or $m = 21\frac{1}{4}$.
- $r \leq 12$ *r* can be any value less than *or* equal to 12.
  Example values: $r = 12$, $r = 9$, or $r = -2.75$.
- $s \geq 15$ *s* can be any value greater than *or* equal to 15.
  Example values: $s = 15$, $s = 20$, or $s = 30\frac{1}{2}$.

Did You Know . . . ?
In an inequality, the allowed values include whole numbers, decimals, and fractions.

For each inequality, circle any allowed value for the letter. Two problems are done as examples.

1. **a.** $n < 9$ (–4) (0) (6) 9 10 12 **b.** $p < 11$ –1 0 4 10 12 14

2. **a.** $t > 10$ –3 1 7 10 13 25 **b.** $x > 7$ –2 0 7 9 18 26

3. **a.** $m \leq 15$ (–3) (1) (15) 18 27 39
   [*m* can be equal to 15 *or* any number less than 15.]
   b. $n \leq 5$ –6 1 4 5 7 10
   [*n* can be equal to 5 *or* any number less than 5.]

4. **a.** $y \geq 12$ –3 1 12 18 27 39
   [*y* can be equal to 12 *or* any number greater than 12.]
   b. $x \geq 8$ –9 4 8 13 18 21
   [*x* can be equal to 8 *or* any number greater than 8.]

Range of Values

Two comparison symbols can be used together to describe a range of values. The lowest end of the range is usually written as the first number of the inequality.

Inequality	Meaning	Example Values
$5 < n < 8$	n is greater than 5 *and* less than 8.	$n = 6$, $n = 7$, $n = 7.5$
$-3 \le m < 2$	m is greater than or equal to −3 *and* less than 2.	$m = -3$, $m = 0$, $m = 1$
$0 < x \le 5$	x is greater than 0 *and* less than or equal to 5.	$x = \frac{3}{4}$, $x = 1$, $x = 4.2$, $x = 5$
$-6 \le y \le 6$	y is greater than or equal to −6 *and* less than or equal to 6.	$y = -6$, $y = 0$, $y = 2.9$, $y = 6$

Write the meaning of each inequality.

5. $-3 < x < 7$ ______________________________

6. $-4 \le y < 20$ ______________________________

7. $6 < p \le 17$ ______________________________

8. $-9 \le x \le 13$ ______________________________

For each inequality, three possible values are given. If the value is correct (makes the inequality a true statement), circle Yes. If not, circle No.

9. $-1 < x < 6$ a. $x = 4$ Yes No   b. $x = -1$ Yes No   c. $x = 0$ Yes No

10. $-3 \le n < 0$ a. $n = 0$ Yes No   b. $n = 1$ Yes No   c. $n = -4$ Yes No

11. $1 < m \le 9$ a. $m = -2$ Yes No   b. $m = 0$ Yes No   c. $m = 5$ Yes No

12. $-7 \le p \le 7$ a. $p = -3$ Yes No   b. $p = 0$ Yes No   c. $p = 3$ Yes No

Graphing an Inequality

An inequality can be graphed on a number line.

- An open circle, ○, means that a number is not a correct value.
- A solid circle, ●, means that a number is a correct value.
- A solid line is drawn through all correct values.

EXAMPLE 1 Inequality: $n > -4$

Values of n

Graph:

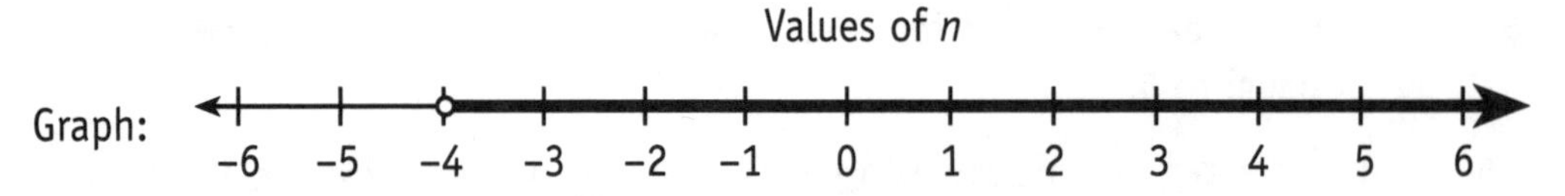

Meaning: n is greater than –4. (n cannot have the value of –4.)

EXAMPLE 2 Inequality: $x \geq 0$

Values of x

Graph:

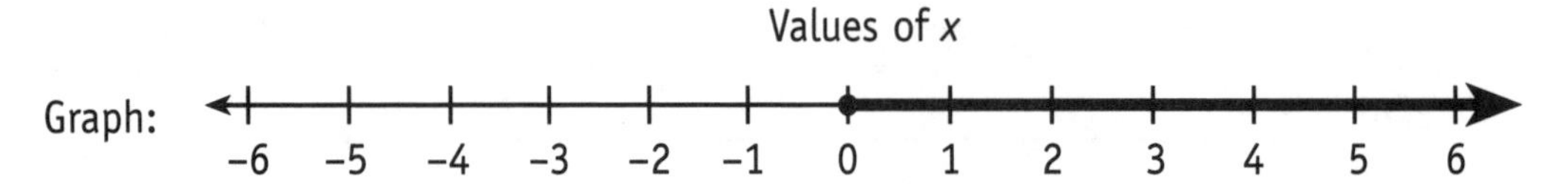

Meaning: x is greater than or equal to 0. (x can have the value 0.)

EXAMPLE 3 Inequality: $-5 < s \leq 4$

Values of s

Graph:

–6 –5 –4 –3 –2 –1 0 1 2 3 4 5 6

Meaning: s is greater than –5 *and* less than or equal to 4.

EXAMPLE 4 Inequality: $-5 < v < 3$

Values of v

Graph:

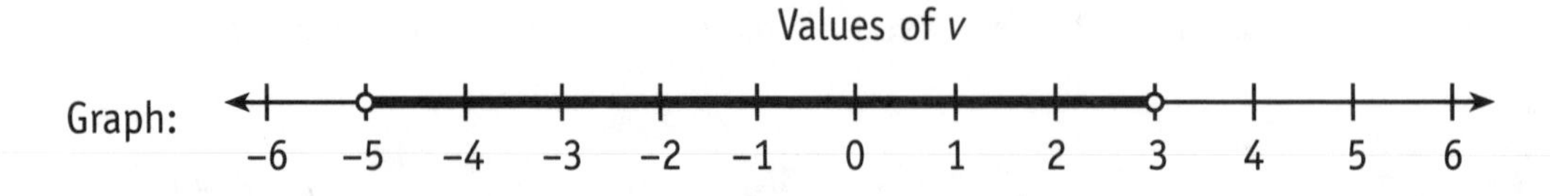

Meaning: v is greater than –5 *and* less than 3.

Choose the inequality that is graphed on each number line.

1. Values of t

−6 −5 −4 −3 −2 −1 0 1 2 3 4 5 6

a. $t < -2$ **b.** $t > -2$ **c.** $t \geq -2$

2. Values of w

−6 −5 −4 −3 −2 −1 0 1 2 3 4 5 6

a. $-5 < w < 1$ **b.** $-5 < w \leq 1$ **c.** $-5 \leq w < 1$

Write the inequality that is graphed on each number line.

3.

4.

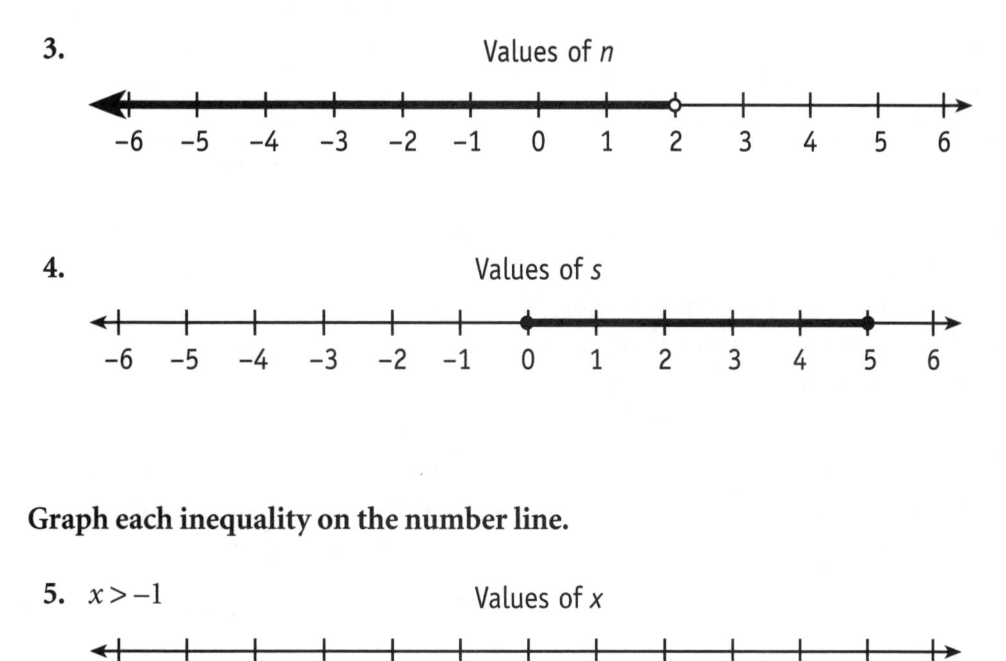

Graph each inequality on the number line.

5. $x > -1$

6. $-3 < y \leq 4$

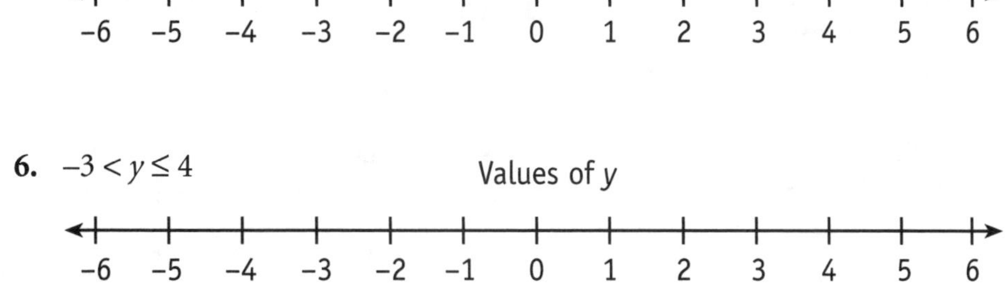

Operations with Numbers

The four basic operations in mathematics are addition, subtraction, multiplication, and division. Addition is always shown by a plus sign (+) and subtraction by a minus sign (–). Multiplication and division can be shown in more than one way.

ADDITION		SUBTRACTION	
Symbol	Example	Symbol	Example
+	4 + 5	–	6 – 3

MULTIPLICATION		DIVISION	
Symbol	Example	Symbol	Example
×	7 × 3	÷	12 ÷ 4
•	7 • 3	/	12/4
( )	7(3) or (7)3	—	$\frac{12}{4}$

Order of Operations

In multistep problems, the order in which you perform operations is important.

- First, perform operations within parentheses.
- Second, multiply or divide, working from left to right.
- Third, add or subtract, working from left to right.

EXAMPLE 1 Simplify 6 + 8 ÷ 4.

STEP 1 Divide first: 8 ÷ 4 = 2

STEP 2 Add: 3 + 2 = **6**

ANSWER: 6

(**Note:** If you add as your first step, you get 14 ÷ 4 = 3.5, which is incorrect.)

EXAMPLE 3 Simplify 12 × 3 – 20.

STEP 1 Multiply first: 12 × 3 = 36

STEP 2 Subtract: 36 – 20 = **16**

ANSWER: 16

EXAMPLE 2 Simplify 3(7– 2).

STEP 1 Subtract the numbers inside the parentheses: 7 – 2 = 5

STEP 2 Multiply: 3(5) = **15**

ANSWER: 15

EXAMPLE 4 Simplify 7(5 + 3) – 10/2.

STEP 1 Add the numbers inside the parentheses: 5 + 3 = 8

STEP 2 Multiply: 7(8) = 56

STEP 3 Divide: 10/2 = 5

STEP 4 Subtract: 56 – 5 = **51**

ANSWER: 51

Simplify each product or quotient.

1.	$5(3) =$	$8 \cdot 3 =$	$(6)3 =$	$5 \times 7 =$	$9 \cdot 2 =$
2.	$30 \div 6 =$	$12/6 =$	$\frac{28}{7} =$	$45 \div 9 =$	$33/11 =$
3.	$9 \cdot 6 =$	$21/7 =$	$4(2) =$	$18 \div 9 =$	$5 \times 8 =$

Show three ways to write each expression.

4. 12 times 9 ______ ______ ______

5. 24 divided by 6 ______ ______ ______

Simplify. The first one in each row is started for you.

6. $5 + 9 \times 2$ → $5 + 18$ $\qquad 12 \times 4 - 6 \qquad 25 - 15 \div 3 \qquad 14 \div 2 + 6$

7. $2(5 + 3)$ → $2(8)$ $\qquad 5(12 - 3) \qquad (25 + 5) \div 6 \qquad (42 - 24) \div 6$

8. $(6 \times 2) + (3 \times 2)$ → $12 + 6$ $\qquad (36 \div 4) - (3 \times 2) \qquad (4 \times 3) - (15 \div 5)$

9. $4(3 + 2) - (3 \times 2)$ → $4(5) - 6$ $\qquad 5(12 - 8) + (5 \times 3) \qquad (12 \div 6) + 2(8 - 5)$

Rounding to a Chosen Place Value

To **round** a number is to simplify its value.

- You round a whole number by replacing one or more digits with zeros.
- You round a decimal number by dropping one or more digits and rounding the remaining digits.

Decimal answers on a calculator often contain more digits than you need, so it is common to round calculator answers.

Rounding Whole Numbers

Whole numbers are most often rounded to the nearest ten, nearest hundred, or nearest thousand.

- Round up when a number is halfway or more between two rounded numbers.
- Round down when a number is less than halfway between two rounded numbers.

Nearest Ten

35 rounds to 40.

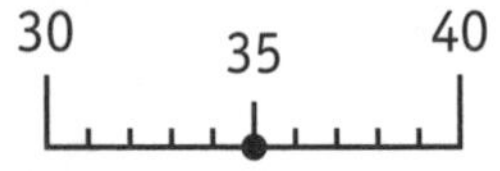

30 35 40

35 is halfway between 30 and 40. 35 rounds up to 40.

Nearest Hundred

128 rounds to 100.

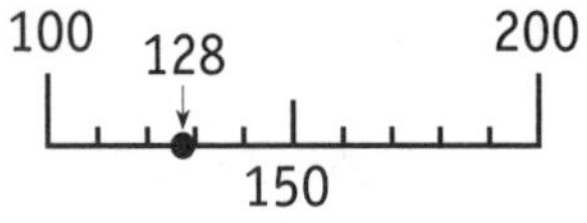

100 128 200
150

128 is less than halfway between 100 and 200. 128 rounds down to 100.

Nearest Thousand

2,685 rounds to 2,000.

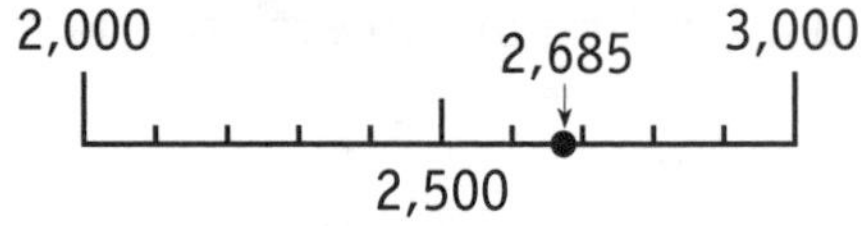

2,000 2,685 3,000
2,500

2,685 is more than halfway between 2,000 and 3,000. 2,685 rounds up to 3,000.

Rounding Decimals

To round a decimal to a chosen place value, look at the digit to the right of the chosen place value.

- If the digit is greater than or equal to 5, round up.
- If the digit is less than 5, do not change the digit in the chosen place value.

When rounding or estimating, the symbol ≈ is often used. ≈ means "is approximately equal to."

Nearest Whole Number

Check the digit in the tenths place.

$3.8 \approx 4$
↑ greater than or equal to 5

$7.2 \approx 7$
↑ less than 5

Nearest Tenth

Check the digit in the hundredths place.

$1.74 \approx 1.7$
↑ less than 5

$9.28 \approx 9.3$
↑ greater than or equal to 5

Nearest Hundredth

Check the digit in the thousandths place.

$2.486 \approx 2.49$
↑ greater than or equal to 5

$5.674 \approx 5.67$
↑ less than 5

Round each number to the nearest ten.

1. 45 ≈ ______ 72 ≈ ______ 18 ≈ ______ 53 ≈ ______

Round each number to the nearest hundred.

2. 150 ≈ ______ 97 ≈ ______ 231 ≈ ______ 479 ≈ ______

Round each number to the nearest thousand.

3. 2,500 ≈ ______ 4,195 ≈ ______ 6,825 ≈ ______ 3,750 ≈ ______

Round each number to the nearest whole number.

4. 7.5 ≈ ______ 8.2 ≈ ______ 11.3 ≈ ______ 16.8 ≈ ______

Round each number to the nearest tenth.

5. 2.65 ≈ ______ 3.37 ≈ ______ 8.72 ≈ ______ 11.66 ≈ ______

Round each number to the nearest hundredth.

6. 1.755 ≈ ______ 4.062 ≈ ______ 2.578 ≈ ______ 5.644 ≈ ______

Round each number as indicated.

7. a sales tax of $0.655

   ______
   (nearest cent)

8. a length of 2.9 m

   ______
   (nearest m)

9. a car price of $11,250

   ______
   (nearest $100)

10. a width of 0.93 cm

    ______
    (nearest 0.1 cm)

11. a house price of $76,825

    ______
    (nearest $1,000)

12. a race of 3.25 miles

    ______
    (nearest 0.1 mile)

Using a Calculator

A hand calculator is a very important math tool. In *Pre-Algebra*, you can practice using a calculator to help solve a wide range of problems.

The Calculator Keyboard

The calculator below is similar to one you've seen or may be using.

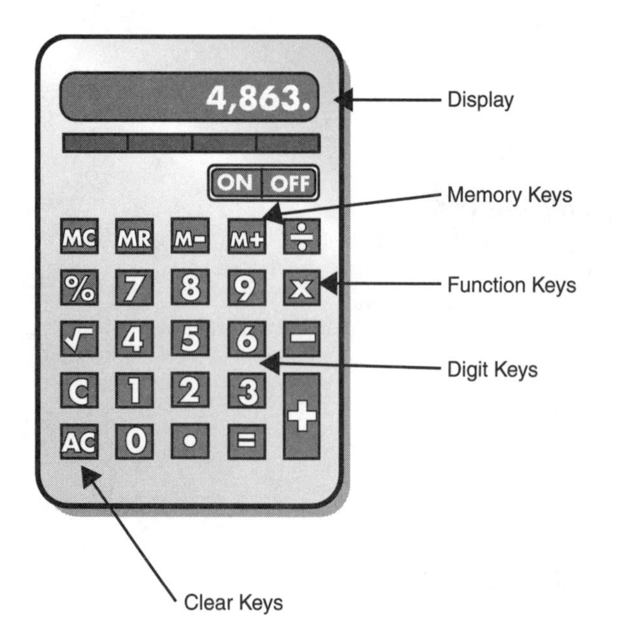

Locate the following (or similar) keys on your calçulator.

- The **on/off key** [ON/OFF]. Press [ON/OFF] once to turn a calculator on and press it again to turn it off.
- The **clear key** [C]. Press [C] to erase the display. Press [C] to begin a new problem or when you've made a keying error.

  Other commonly used clear key symbols are:

  [ON/C] On/Clear [CE/C] Clear Entry/Clear [CE] Clear Entry [AC] All Clear
- The **digits keys** [0], [1], [2], [3], [4], [5], [6], [7], [8], and [9]. To enter a number, press one digit at a time.
- The **decimal point key** [·]. Use this key to enter a decimal point in a decimal or to separate dollars from cents in a money amount.
- The **function keys** [+], [−], [×], and [÷]. Press a function key to add, subtract, multiply, or divide.
- The **equals key** [=]. Press [=] to display the answer to a calculation.

Entering Numbers on a Calculator

To enter a number on a calculator, press one digit key at a time. On the display on the previous page, the number 4,863 is entered. Notice the following features.

- A decimal point is displayed to the right of the ones digit.
- A comma separates groups of digits. However, not all calculators display a comma.
- The calculator does *not* have a comma (,) key or a dollar sign ($) key.

Try the next two examples on your calculator.

EXAMPLE 1 Divide 351 by 13.

STEP 1 Press the clear key to clear the display.

STEP 2 Press the digit keys, divide key (÷), and equals key (=).

Press Keys	Display Shows
[C] [3] [5] [1] [÷] [1] [3] [=]	27.

ANSWER: 27

EXAMPLE 2 Multiply $16.35 by 8.

STEP 1 Press the clear key to clear the display.

STEP 2 Press the digit keys, decimal point key, times key (×), and equals key (=).

Press Keys	Display Shows
[C] [1] [6] [·] [3] [5] [×] [8] [=]	130.8

ANSWER: $130.80

↑ Press the decimal point key to separate dollars from cents.

↑ Calculators do not show zeros to the right of a decimal fraction.

Use a calculator to solve the following problems.

1. $246 + 159 + 98$

2. $2{,}456 - 897$

3. $2{,}258 + 1{,}946 + 958$

4. $3{,}897 - 1{,}879$

5. 148×29

6. $8\overline{)1{,}992}$

7. 419(326)

8. $30,430 ÷ 34

Terminating and Repeating Decimals

A decimal results each time you

- divide one number by another and get a remainder
  EXAMPLE $7 \div 4 = 1.75$
- divide a smaller number by a larger number
  EXAMPLE $6 \div 9 = 0.666\ldots = 0.\overline{6}$ (The small bar tells you that the 6's don't ever end!)

For each type of division, the decimal that results is called either a terminating decimal or repeating (non-terminating) decimal. The examples below will show how the two types of decimal answers appear on a calculator.

Terminating Decimal

A terminating decimal has a limited number of digits. Most division problems have four or fewer digits to the right of the decimal point.

EXAMPLES

	Press Keys	Display Shows
One decimal digit: $4 \div 5 = 0.8$	C 4 ÷ 5 =	0.8
Two decimal digits: $3 \div 4 = 0.75$	C 3 ÷ 4 =	0.75
Three decimal digits: $5 \div 8 = 0.625$	C 5 ÷ 8 =	0.625

Repeating Decimal

A repeating decimal has a never-ending, repeating pattern of one or more digits. A repeating decimal is often written with a bar placed over the repeating digit(s).

EXAMPLES

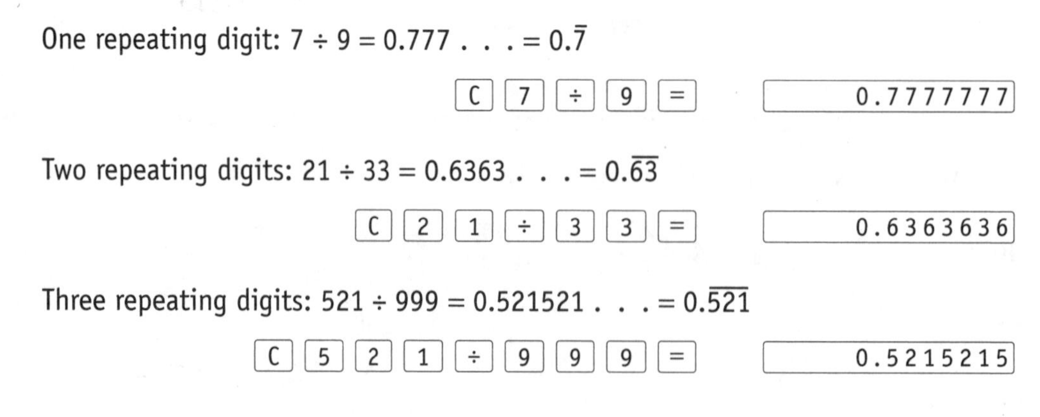

Rounding versus Truncating

Notice that a repeating digit continues forever! For this reason, a calculator gives an approximate answer to any problem that results in a repeating decimal. (**Note:** Most calculators display a maximum of eight digits.)

Try this division problem on your calculator: 2 ÷ 3.

- If the answer displayed is 0.6666666, your calculator **truncates** (drops) any digits that won't fit on the display.
- If the answer displayed is 0.6666667, your calculator **rounds** the answer to the final decimal place on your calculator display.

Most inexpensive calculators truncate. Many expensive calculators—those designed for use in business and in science—round repeating decimals.

Divide, using pencil and paper. Check (✓) the type of decimal answer you obtain.

1. 5)3 ______ terminating
______ repeating

2. 11)3 ______ terminating
______ repeating

3. 6)2 ______ terminating
______ repeating

4. 8)5 ______ terminating
______ repeating

Divide, using a calculator. Round each answer to the thousandth place. Problems 5 and 7 are done as examples.

5. 11 ÷ 9 = [1.2222222] ≈ **1.222**

6. 4 ÷ 3 = [ ] ≈

7. 15 ÷ 33 = [0.4545454] ≈ **0.455**

8. 4 ÷ 11 = [ ] ≈

9. 208 ÷ 333 = [ ] ≈

10. 15 ÷ 111 = [ ] ≈

Some answers neither terminate nor show their repeating pattern in the first eight digits. Write the displayed answers to the following quotients.

11. 3 ÷ 7 = [ ]

12. 5 ÷ 17 = [ ]

Number Skills Review

Work each problem and check your answers. Correct any errors.

1. What fraction of the circle is shaded?

2. What part of the total bar is shaded? Write your answer as a decimal.

3. What percent of the grid is shaded?

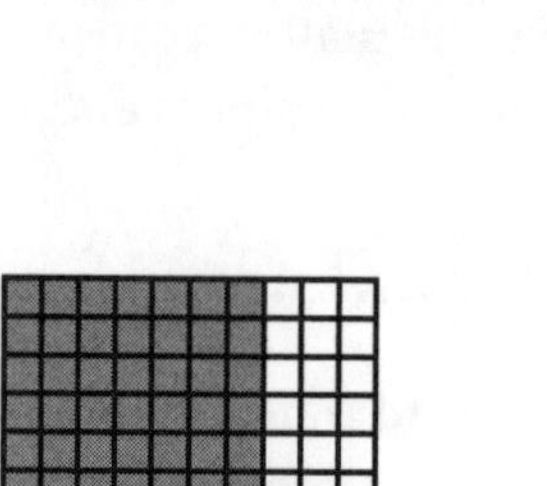

4. Write 35% as a fraction and as a decimal.

   fraction: ______

   decimal: ______

Solve each problem.

5. $\frac{2}{3} + \frac{1}{6}$ $\qquad$ $\frac{3}{4} \div \frac{2}{5} =$ $\qquad$ 3.6×0.05 $\qquad$ 15% of 90 =

6. On a number line, what number is 5 units to the left of the number 3?

7. In words, what does the inequality $n \geq 7$ mean?

8. In words, what does the inequality $-2 \leq x < 1$ mean?

Write the inequality that is graphed on each number line.

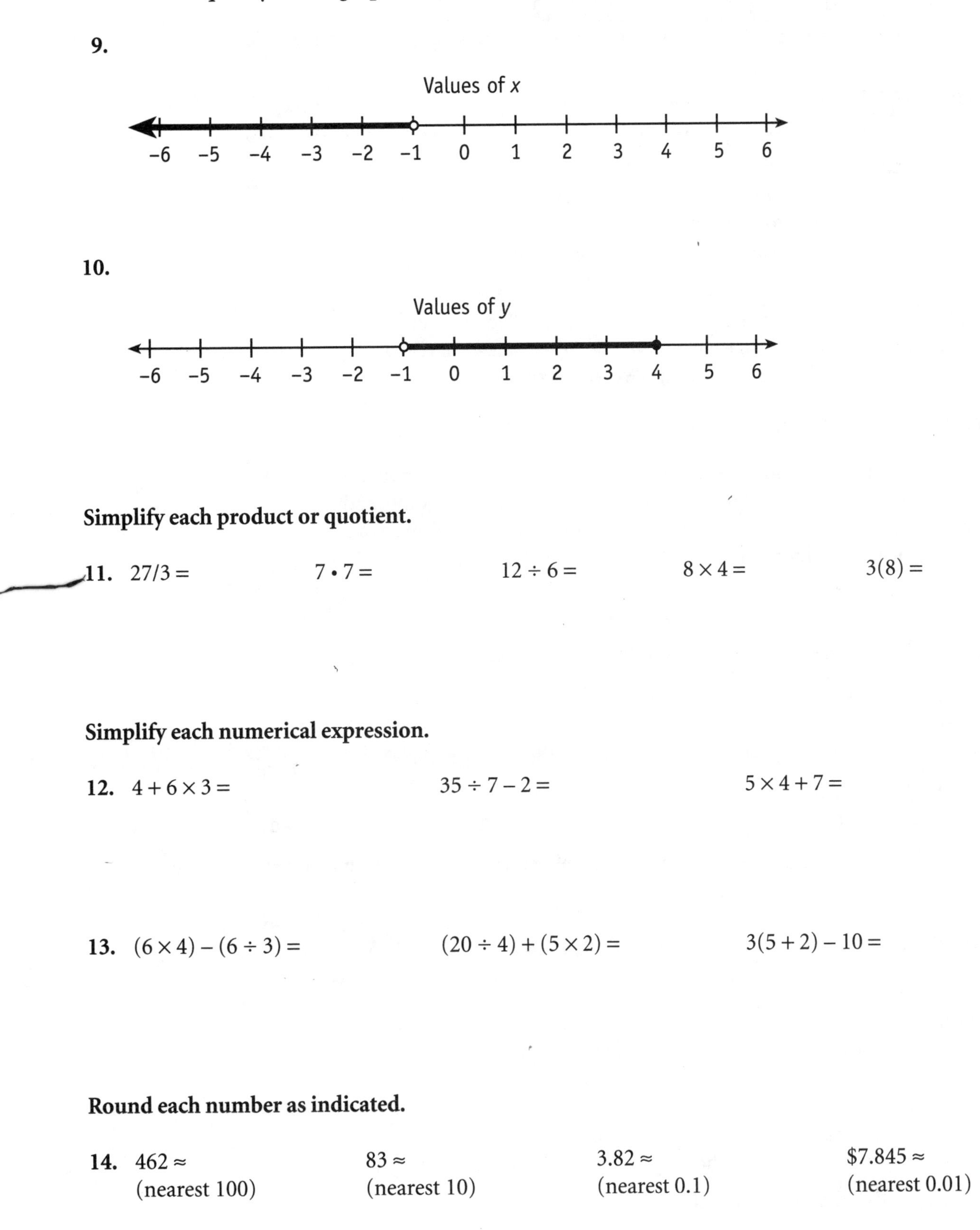

9.

10.

Simplify each product or quotient.

11. $27/3 =$ $\quad 7 \cdot 7 =$ $\quad 12 \div 6 =$ $\quad 8 \times 4 =$ $\quad 3(8) =$

Simplify each numerical expression.

12. $4 + 6 \times 3 =$ $\quad 35 \div 7 - 2 =$ $\quad 5 \times 4 + 7 =$

13. $(6 \times 4) - (6 \div 3) =$ $\quad (20 \div 4) + (5 \times 2) =$ $\quad 3(5 + 2) - 10 =$

Round each number as indicated.

14. $462 \approx$ (nearest 100) $\quad 83 \approx$ (nearest 10) $\quad 3.82 \approx$ (nearest 0.1) $\quad \$7.845 \approx$ (nearest 0.01)

WORD PROBLEM SKILLS

Estimating

To **estimate** is to find an approximate answer. You estimate to

- check addition, subtraction, multiplication, or division
- find an approximate answer
- choose an answer in a multiple-choice question

To estimate, use rounded numbers, numbers that are easy to work with. Then add, subtract, multiply, or divide as needed. A careful estimate will be close to an exact answer. On tests, an estimate may enable you to quickly decide among several answer choices.

Estimating with Whole Numbers

To estimate when adding, subtracting, or multiplying with whole numbers, round each number to the nearest ten, hundred, or thousand—depending on the size of the numbers. To estimate when dividing, round to **compatible numbers** (numbers that are easily divided).

EXAMPLE 1

Problem	Estimate
$\begin{array}{r} 415 \\ +294 \\ \hline \end{array}$	$\begin{array}{r} 400 \\ +300 \\ \hline \mathbf{700} \end{array}$

STEP 1 Round each number to the nearest hundred.

STEP 2 Add 400 plus 300.

ESTIMATE: 700
ANSWER: 709

EXAMPLE 2

Problem	Estimate
$\begin{array}{r} 39 \\ \times 23 \\ \hline \end{array}$	$\begin{array}{r} 40 \\ \times 20 \\ \hline \mathbf{800} \end{array}$

STEP 1 Round each number to the nearest ten.

STEP 2 Multiply 40 times 20.

ESTIMATE: 800
ANSWER: 897

EXAMPLE 3

Problem	Estimate
$9\overline{)357}$	$\begin{array}{r} \mathbf{40} \\ 9\overline{)360} \end{array}$

STEP 1 Round 357 to 360. (9 and 360 are compatible numbers: 9 divides evenly into 360.)

STEP 2 Divide 360 by 9.

ESTIMATE: 40
ANSWER: 39 r 6

Estimate each answer. Then use your estimate to choose the exact answer. Finally, calculate the exact answer to show that you are correct.

1. In 1984 the population of Oakville was 39,877. Today the population is 50,316. By how much has the population of Oakville increased since 1984?

   a. 8,649
   b. 8,899
   c. 9,229
   d. 10,439
   e. 11,789

   Estimate ____________

   Exact Answer ____________

2. On June 30 the mileage indicator on Manuel's car read 5,148. In July Manuel drove a total of 1,682 miles. In August he drove 1,209 miles. By the end of August, how many miles had Manuel's car been driven?

a. 6,549
b. 6,929
c. 8,039
d. 9,499
e. 9,909

Estimate ____________

Exact Answer ____________

3. Leng's car gets 43 miles to the gallon. At this rate, how many miles can Leng drive with 18 gallons of gasoline?

a. 714
b. 774
c. 864
d. 914
e. 984

Estimate ____________

Exact Answer ____________

4. Jani baked 231 cookies. She kept 48 for herself and divided the rest equally among three fourth grade classes. How many cookies did each class get?

a. 49
b. 53
c. 61
d. 68
e. 72

Estimate ____________

Exact Answer ____________

5. Sky View Cinema sold 197 large cola drinks Saturday. If each large cup holds 21 fluid ounces, how many total fluid ounces of cola were sold Saturday?

a. 2,247
b. 2,977
c. 3,387
d. 3,657
e. 4,137

Estimate ____________

Exact Answer ____________

6. Xiaoyu cut a rope into four equal parts. If the uncut rope was 38 feet 8 inches long, how long is each cut piece?

a. 9 feet 8 inches
b. 9 feet 1 inch
c. 8 feet 7 inches
d. 8 feet 2 inches
e. 7 feet 11 inches

Estimate ____________

Exact Answer ____________

Estimating with Decimals

To estimate when adding, subtracting, or multiplying decimals, round each decimal to the nearest whole number. To estimate when dividing, round to **compatible numbers** (numbers that are easily divided).

EXAMPLE 1

Problem	Estimate
7.85 − 1.96	8 − 2 = **6**

STEP 1 Round each decimal to the nearest whole number.

STEP 2 Subtract 2 from 8.

ESTIMATE: 6
ANSWER: 5.89

EXAMPLE 2

Problem	Estimate
12.2 × 5.9	12 × 6 = **72**

STEP 1 Round each decimal to the nearest whole number.

STEP 2 Multiply 12 times 6.

ESTIMATE: 72
ANSWER: 71.98

EXAMPLE 3

Problem	Estimate
$4\overline{)26.92}$	$4\overline{)28}$ = **7**

STEP 1 Round 26.92 to 28. (4 and 28 are compatible numbers: 4 divides evenly into 28.)

STEP 2 Divide 28 by 4.

ESTIMATE: 7
ANSWER: 6.73

Estimate each answer. Then use your estimate to choose the exact answer. Finally, calculate the exact answer to show that you are correct.

7. Lucille paid $129.75 for a lawnmower. If she paid by cashing a paycheck written for $182.90, how much change should she receive?

   a. $38.85
   b. $42.25
   c. $53.15
   d. $61.95
   e. $67.45

   Estimate 50.00

   Exact Answer 53.15

8. At a price of $4.89 per pound, what is the cost of 3.95 pounds of deluxe mixed nuts?

   a. $19.32
   b. $18.12
   c. $17.52
   d. $16.92
   e. $15.72

   Estimate 20.00

   Exact Answer 19.32

9. A ribbon is 147.6 centimeters long. If Kristen wants to cut the ribbon into three equal pieces, how long should she cut each piece?

a. 36.6 cm
b. 39.3 cm
c. 43.4 cm
d. 49.2 cm
e. 54.9 cm

Estimate ____________

Exact Answer ____________

10. Pixie is mailing three presents inside a box. The presents weigh 1.06 kilograms, 2.85 kilograms, and 4.13 kilograms. If the box weighs 1.2 kilograms, what is the total weight of the box with the presents inside?

a. 7.52 kg
b. 8.12 kg
c. 9.24 kg
d. 11.32 kg
e. 12.62 kg

Estimate ____________

Exact Answer ____________

11. On Saturday Maria sold three loads of topsoil at her garden shop. The weights of the loads were 1,340 pounds; 1,790 pounds; and 1,412 pounds. What total weight of topsoil did Maria sell?

a. 4,542 lb
b. 4,212 lb
c. 3,982 lb
d. 3,672 lb
e. 3,452 lb

Estimate ____________

Exact Answer ____________

12. Bryan paid $68.75 for three shirts. The prices of the shirts were $18.95, $31.14, and $14.85. How much change did Bryan receive?

a. $3.81
b. $8.28
c. $7.19
d. $1.12
e. $9.06

Estimate ____________

Exact Answer ____________

Estimating with Mixed Numbers

To estimate when adding, subtracting, multiplying, or dividing with mixed numbers, round each mixed number to the nearest whole number.

EXAMPLE 1

Problem	Estimate
$5\frac{7}{8}$	6
$+\,3\frac{1}{4}$	$+\,3$
	9

STEP 1 Round each mixed number to the nearest whole number.

STEP 2 Add 6 plus 3.

ESTIMATE: 9

ANSWER: $9\frac{1}{8}$

EXAMPLE 2

Problem	Estimate
$4\frac{1}{3}$	4
$\times\,1\frac{3}{4}$	$\times\,2$
	8

STEP 1 Round each mixed number to the nearest whole number.

STEP 2 Multiply 4 times 2.

ESTIMATE: 8

ANSWER: $7\frac{7}{12}$

EXAMPLE 3

Problem	Estimate
$11\frac{1}{3} \div 2\frac{1}{4}$	$12 \div 2 = \mathbf{6}$

STEP 1 Round $11\frac{1}{3}$ to 12 and round $2\frac{1}{4}$ to 2. (12 and 2 are compatible numbers: 2 divides evenly into 12.)

STEP 2 Divide 12 by 2.

ESTIMATE: 6

ANSWER: $5\frac{1}{27}$

Estimate each answer. Then use your estimate to choose the exact answer. Finally, calculate the exact answer to show that you are correct.

13. Ephran mixed $2\frac{1}{4}$ pounds of peanuts with $2\frac{7}{8}$ pounds of deluxe mixed nuts. How many total pounds of nuts does Ephran have in the new mixture?

a. $3\frac{3}{4}$ lb **c.** $5\frac{1}{8}$ lb
b. $4\frac{1}{2}$ lb **d.** $5\frac{1}{2}$ lb

Estimate 5

Exact Answer 5 1/8

14. Jan's pickup can carry $1\frac{3}{4}$ tons of gravel in one load. How many tons of gravel can Jan haul in 11 loads?

a. $12\frac{3}{4}$ T **c.** $17\frac{1}{2}$ T
b. $15\frac{3}{4}$ T **d.** $19\frac{1}{4}$ T

Estimate 20

Exact Answer 19 1/4

15. Justin has a piece of chain that is $28\frac{3}{4}$ feet long. If he divides the chain into five equal pieces, how many feet long will each shorter piece be?

a. $5\frac{3}{4}$ ft   **c.** $7\frac{3}{4}$ ft

b. $9\frac{1}{2}$ ft   **d.** $8\frac{1}{4}$ ft

Estimate ____________

Exact Answer ____________

16. Before she made curtains, Caren had $26\frac{1}{3}$ yards of fabric. She now has $12\frac{2}{3}$ yards. How many yards of material did Caren use for the curtains?

a. $11\frac{1}{3}$ yd   **c.** $15\frac{1}{3}$ yd

b. $13\frac{2}{3}$ yd   **d.** $16\frac{2}{3}$ yd

Estimate ____________

Exact Answer ____________

17. To cover a living room wall, Ernesto used $7\frac{1}{4}$ strips of wallpaper. If each strip is $1\frac{3}{4}$ feet wide, how many feet wide is the wall that Ernesto covered?

a. $8\frac{3}{4}$ ft   **c.** $11\frac{1}{4}$ ft

b. $9\frac{7}{8}$ ft   **d.** $12\frac{11}{16}$ ft

Estimate ____________

Exact Answer ____________

18. Wendy uses $1\frac{7}{8}$ ounces of silver for each buckle she makes to sell at a crafts fair. How many complete buckles can Wendy make from her remaining $20\frac{1}{8}$ ounces of silver? (**Note:** The answer is a whole number.)

a. 7   **c.** 10

b. 8   **d.** 13

Estimate ____________

Exact Answer ____________

Understanding Set-Up Questions

A **set-up question** asks you to choose an expression for a calculation not yet completed that shows *how to find a correct answer.* In a set-up question, you are not asked to find the answer itself! Two kinds of set-up questions are shown in the examples below.

EXAMPLE 1 Kelly has run 7.8 miles of a 10-mile race. Which numerical expression tells how far Kelly still has left to run?

a. $10 + 7.8$ **c.** $7.8 - 10$ **e.** 10×7.8
b. $7.8 + 10$ **d.** $10 - 7.8$

Two or more numbers combined by +, −, ×, or ÷ signs is called a **numerical expression.**

You think: How much is 10 subtract 7.8?

Choose the expression that shows this subtraction.

ANSWER: d. 10 − 7.8

EXAMPLE 2 At a holiday sale, the price of flowers is being reduced by 30%. Which numerical expression tells the amount you save (s) when buying a bouquet of roses that normally sells for $14.00?

a. $s = 0.03 \times \$14.00$ **c.** $s = 0.03 \div \$14.00$ **e.** $s = 0.3 \div \$14.00$
b. $s = 0.3 \times \$14.00$ **d.** $s = \$14.00 \div 0.3$

You think: How do I find 30% of $14.00?

Choose the expression that shows a savings (s) equal to $0.3 \times \$14.00$.

ANSWER: b. $s = 0.3 \times \$14.00$

Write a numerical expression for each amount. The first problem is done as an example.

Amount	Numerical Expression
1. $74 subtract $28	$74 – $28
2. the product of $5.75 times 6	
3. 368 divided by 6	
4. 35% of $90	
5. the sum of 147 and 89	
6. the difference between $2\frac{1}{2}$ and $1\frac{3}{4}$	
7. 20% of 164	
8. $18.60 divided equally four ways	

Choose the correct numerical expression that solves each problem.

9. Nick poured 0.8 liters of juice out of a bottle that holds 2.5 liters. Which expression tells how many liters of juice is left?

a. $8 - 2.5$ **b.** $0.8 + 2.5$ **c.** $2.5 - 0.8$

10. In Scott's pocket are two quarters, three dimes, and one nickel. Which expression tells how much money (m) Scott has?

a. $m = 50¢ + 30¢ + 5¢$ **b.** $m = 25¢ + 30¢ + 10¢$ **c.** $m = 20¢ + 30¢ + 10¢$

11. Lucy cut an 18-foot ribbon into four equal parts. Which expression gives the length, in feet, of each part?

a. $18 - 4$ **b.** $18 \div 4$ **c.** $4 \div 18$

12. To make light blue paint, Tom mixes 3 fluid ounces of blue tint with 1 gallon of white paint. Which expression tells how much tint Tom should mix with $\frac{1}{4}$ gallon of white paint to make light blue paint?

a. 3×4 **b.** 3×1.25 **c.** 3×0.25

13. When Jeff bought a tennis racket, he received $17.50 in change. If the racket cost $62.50, which expression tells how much Jeff paid the salesperson?

a. $62.50 – $17.50 **b.** $62.50 + $17.50 **c.** $80.00 – $62.50

14. At a 40% off sale, Marnie bought a sweater. If the original price of the sweater was $48, which expression tells how much Marnie saved?

a. $\frac{3}{5} \times \$48$ **b.** $\frac{5}{2} \times \$48$ **c.** $\frac{2}{5} \times \$48$

15. Fifteen of the 20 students in Sally's art class are women. Which expression can be used to find the percent of the students (p) who are men?

a. $p = \frac{1}{4} \times 100\%$ **b.** $p = 100\% \div \frac{3}{4}$ **c.** $p = \frac{3}{4} \times 100\%$

16. Blake paid $82.75 for a stereo that usually costs $118.79. Which expression gives the *best estimate* of how much Blake saved?

a. $110 – $80 **b.** $120 – $90 **c.** $120 – $80

17. A recipe calls for $\frac{3}{4}$ cup of sugar. Laura plans to make $2\frac{1}{2}$ times the number of cookies in the recipe. Which expression tells how many cups of sugar (c) is needed?

a. $c = \frac{3}{4} \times \frac{5}{2}$ **b.** $c = \frac{3}{4} \times \frac{2}{4}$ **c.** $c = \frac{3}{4} \times \frac{5}{4}$

Set-Up Questions in Multistep Problems

Set-up questions can also be written for problems that take more than one step to solve.

EXAMPLE For school Erin bought six pens for $1.25 each and she bought a folder for $3.50. Which numerical expression shows how much Erin paid for these supplies?

a. $1.25 + $3.50 + $6.00
b. ($3.50 × 6) + $1.25
c. ($3.50 – $1.25) × 6
d. ($1.25 × 6) + $3.50
e. ($1.25 + $3.50) × 6

Erin paid $7.50 ($1.25 × 6) for the pens, and she paid $3.50 for the folder. Her total cost is $11.00 ($7.50 + $3.50).

The correct answer choice is **d. ($1.25 × 6) + $3.50.**

None of the other answer choices gives $11.00 as an answer.

a. $1.25 + $3.50 + $6.00 = $10.75
b. ($3.50 × 6) + $1.25 = $21.00 + $1.25 = $22.25
c. ($3.50 – $1.25) × 6 = $2.25 × 6 = $13.50
e. ($1.25 + $3.50) × 6 = $4.75 × 6 = $27.00

Remember:
Do calculations inside parentheses as your first step.

Write a numerical expression for each amount. The first problem is done as an example.

	Amount	Numerical Expression
1.	5 times the sum of 27 plus 14	5(27+ 14)
2.	$24.75 added to the product of $6.50 times 9	________
3.	$10 subtracted from the quotient of $150 divided by 4	________
4.	30% of $60 added to the sum of $15 plus $12	________
5.	6 times the difference of $13.50 subtract $4.25	________
6.	15 added to the product of 13 times 2	________
7.	20% of 24 subtracted from 30	________
8.	dividing the sum of $24.50 plus $12.00 six ways	________

Choose the correct numerical expression that solves each problem.

9. Shauna and two friends bought a microwave oven for \$165. Three weeks later, they received a \$20 rebate check from the manufacturer. If they share all costs equally, which expression shows each person's share after dividing the rebate?

a. ($165 ÷ 3) + $20
b. ($165 ÷ 3) – $20
c. ($165 – $20) ÷ 3
d. ($165 + $20) ÷ 3

10. John practices piano for 30 minutes Monday through Friday. On Saturday he practices for 45 minutes. Which expression tells how many minutes John practices piano during these 6 days?

a. (30 × 6) + 45
b. (30 + 45) × 6
c. (45 × 6) + 30
d. (30 × 5) + 45

11. Janice bought three blouses on sale for \$21.95 each. She paid the clerk with four \$20 bills. Which expression tells how much Janice should receive in change?

a. ($21.95 × 3) – ($20.00 × 4)
b. ($20.00 × 4) – ($21.95 × 3)
c. ($21.95 × 4) – ($20.00 × 3)
d. ($21.95 × 3) + ($20.00 × 4)
e. ($20.00 × 4) + ($21.95 × 3)

12. James has 1.5 cases of motor oil in his garage. Each case contains 12 liters of oil. Each time James changes oil, he uses 3 liters of oil. Which expression tells how many times James can change oil before his supply is gone?

a. (12 + 13) ÷ 1.5
b. (12 + 1.5) ÷ 3
c. (12 – 1.5) ÷ 3
d. (12 × 3) ÷ 1.5
e. (12 × 1.5) ÷ 3

13. Miranda has 256 quarters in her coin collection. She kept 125 quarters and divided the rest between her two best friends. Which expression tells how many quarters she gave each friend?

a. (256 – 125) ÷ 2
b. (125 – 256) ÷ 3
c. (256 + 125) ÷ 2
d. (256 – 125) ÷ 3
e. (256 + 125) ÷ 2

14. Stacey and three friends have agreed to share equally the cost of dinner. They had a pizza for \$13.50 and soft drinks for \$3.80. What is Stacey's share of the dinner bill?

a. ($13.50 – $3.80) ÷ 4
b. ($13.50 – $3.80) ÷ 3
c. ($13.50 + $3.80) ÷ 4
d. $13.50 + ($3.80 ÷ 4)
e. $13.50 – ($3.80 ÷ 3)

Finding an Average

Questions often ask you to find an **average.** To find the average of a group of numbers, first find the sum. Then divide the sum by the number of numbers in the group. The average represents a *typical amount* but is often *not* equal to any number in the group.

EXAMPLE Four students took a math quiz on Thursday and got the following scores: 92, 88, 85, and 83. What was the average score?

STEP 1 Add the four scores.

$$\begin{array}{r} 92 \\ 88 \\ 85 \\ +\ 83 \\ \hline 348 \end{array}$$

STEP 2 Divide 348 by 4.

$$\begin{array}{l} \quad\ \mathbf{87} \leftarrow \text{average score} \\ 4\overline{)348} \leftarrow \text{sum of scores} \\ \uparrow \\ \text{number of scores} \end{array}$$

ANSWER: The average score was 87.

> **Do You See . . . ?**
> The average score, 87, is not equal to any of the four scores received by the students.

Solve each problem.

1. On three practice math tests, Licia scored 71%, 82%, and 78%. What was Licia's average percent score on these three tests?

2. Stan bought two shirts on Saturday. One shirt cost $18.00 and the other cost $13.50. What was the average price Stan paid for the shirts?

3. During the first five games of the season, the Rockets scored 84, 92, 70, 78, and 96 points. What was the Rocket's average score for these five games?

4. In her first four 100-meter races, Basha had times of 12.25 seconds, 11.80 seconds, 12.48 seconds, and 11.95 seconds. What is Basha's average time for these four races?

5. Betty rented two movies over the weekend. The first lasted 2 hours 10 minutes and the second lasted 1 hour 50 minutes. What is the average length of these movies?

6. Jason caught four fish. The fish had lengths of 8 inches, $9\frac{1}{2}$ inches, $7\frac{1}{4}$ inches, and $8\frac{3}{4}$ inches. What was the average length of the fish Jason caught?

Identifying Missing Information

Sometimes, a multiple-choice question does not give all the information you need. In fact, a question may ask you to identify information that is missing!

EXAMPLE Alyce and four friends are going to share equally the price of a pizza dinner and several sodas. They ordered two large pizzas and five sodas. Each pizza cost $13.00. How much is Alyce's share of the bill?

a. $4.50
b. $5.20
c. $6.30
d. $6.50
e. not enough information given

The correct answer choice is **e. not enough information given.** To determine Alyce's share of the bill, *you must first know the total bill!* You can't figure out the total bill unless you first know the cost of the sodas. The cost of the sodas is not given.

Being able to identify missing information is an important math skill.

Identify the information that is needed for you to solve each problem. The first problem is done as an example.

1. Richie cut a ribbon into four equal pieces. What is the length of each cut piece?

   the length of the ribbon before it was cut

2. When the price of salmon was reduced by 25% per pound, Keisha bought 1 pound. How much did Keisha pay for the salmon?

3. To make punch, Sean places 0.75 liters of orange juice in a container that holds 2 liters. He then adds 0.5 liters of lemon-lime soda and a small glass full of pineapple juice. What total amount of punch did Sean make?

4. Lynn's basketball team lost Saturday's game by 14 points. How many points did her team score?

5. How many calories are in a meal that includes a 425-calorie hamburger, a medium soft drink, and a 300-calorie order of french fries?

6. Justin is mailing three packages. The large package weighs 7 pounds. The small package weighs 2.5 pounds. What is the average weight of the three packages?

Understanding Ratios

A **ratio** is a comparison of two numbers. For example, if there are 4 apples and 3 oranges in a bowl of fruit, the ratio of apples to oranges is 4 to 3. In words, a ratio is read with the word *to.*

You can write the ratio *4 to 3* in two ways.

- as a fraction, *4 to 3* is $\frac{4}{3}$
- using a colon, *4 to 3* is 4:3

Read the ratios $\frac{4}{3}$ and 4:3 as "four to three."

Write a ratio in the same order as its terms appear in a question. In the example above, the ratio of oranges to apples is 3 to 4.

To simplify a ratio, write it as a fraction and follow these three rules.

1. Reduce a ratio to lowest terms.

   $6 \text{ to } 12 = \frac{6}{12} = \frac{1}{2}$

2. Leave an improper-fraction ratio as an improper fraction.

   $10 \text{ to } 8 = \frac{10}{8} = \frac{5}{4}$

3. Write a whole-number ratio as an improper fraction.

   $3 \text{ to } 1 = \frac{3}{1}$

Simplify each ratio by reducing it to lowest terms.

1.	4 to 8	6 to 9	2 to 8	5 to 15	20 to 50
2.	12 to 4	9 to 6	14 to 7	10 to 8	12 to 10

Solve each ratio problem.

3. Jesse's basketball team won 6 games and lost 4. What is the ratio of *games won to games lost?*

4. In Mariah's kindergarten class, there are 9 girls and 15 boys.

a. What is the ratio of *girls to boys?*

b. What is the ratio of *boys to girls?*

5. Willie weighs 120 pounds and his brother weighs only 80 pounds. What is the ratio of *Willie's weight to his brother's weight?*

6. One quart contains 32 fluid ounces. One cup contains 8 fluid ounces.

a. What is the ratio of *1 quart to 3 cups?*

b. What is the ratio of *1 cup to 2 quarts?*

A ratio problem may involve two steps.

EXAMPLE Only 8 of the 24 students in Lydia's art class are men. In this class, what is the ratio of men to women?

STEP 1 Determine how many women are in the class.

Number of women = 24 − 8 = 16

STEP 2 Write a ratio of *men to women.*

Men to women = $\frac{\text{men}}{\text{women}} = \frac{8}{16} = \frac{1}{2}$

Solve each ratio problem.

7. Last summer Stacey's softball team won 6 of the 9 games they played.

   a. What is the ratio of the games they won to the games they played?

   b. What is the ratio of the games they won to the games they lost?

8. Ms. James has 14 girls and 18 boys in her first grade class.

   a. What is the ratio of girls to boys?

   b. What is the ratio of girls to the total number of students?

   c. What is the ratio of boys to the total number of students?

9. For the after-school barbecue, 12 parents brought main dishes, 6 parents brought desserts, 4 parents brought salads, and 5 parents brought beverages.

   a. What is the ratio of the number of parents who brought salads to the number who brought main dishes?

   b. What is the ratio of the number of parents who brought main dishes to the total number of parents who brought food or beverages?

10. Of the 18 teachers in Shane's school, 14 are women. What is the ratio of men teachers to women teachers at this school?

For problem 11, refer to the drawing at the right.

11. What is the ratio of the distance between Franklin and Bend to the distance between Franklin and Parker?

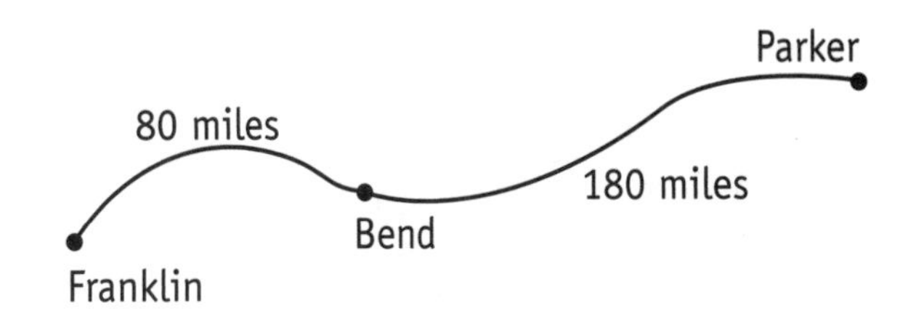

Working with Rates

A **rate** is a ratio that compares two types of measurements. A rate is usually given as a fraction with a denominator of 1. The word *per* is used with rates. Finding a rate can be used as the first step in solving many types of problems.

Are you familiar with these common rates?

cost *per* unit
miles *per* hour
dollars *per* pound
words *per* minute
calories *per* ounce
miles *per* gallon
beats *per* minute
pages *per* minute

EXAMPLE A car travels 117 miles in 3 hours. At this rate, how far will the car travel in 5 hours?

STEP 1 Divide to find the rate in miles *per* hour that the car is traveling. A common name for this rate is *speed*.

Rate = $\frac{117 \text{ miles}}{3 \text{ hours}} = \frac{39 \text{ miles}}{1 \text{ hour}}$ *or* 39 miles *per* hour

STEP 2 To find out how far the car will travel in 5 hours, multiply the rate (39) by 5.

distance = rate × time = 39 × 5 = **195 miles**

ANSWER: 195 miles

Solve each problem.

1. A brand of vitamin pills comes in two sizes. The smaller bottle of 60 tablets costs $5.40. The larger bottle of 90 tablets costs $7.20. Find the cost per tablet (rate) for each bottle. Circle the better buy.

   small bottle:__________ large bottle:__________

2. Manuel's car traveled 66 miles on 3 gallons of gasoline. At this rate, how far can Manuel drive on 19 gallons?

   STEP 1 Find the rate in *miles per gallon* that Manuel's car gets. A common name for this rate is *mileage.*

   STEP 2 Multiply the rate by 19, the number of gallons in Manuel's car.

3. Lam walks 9 miles in 2 hours. At this rate, how far can Lam walk in 7 hours?

STEP 1 Find the rate in *miles per hour* that Lam walks.

STEP 2 Multiply the rate by 7, the number of hours Lam walks.

Did You Know . . . ?
Sometimes a rate is a mixed number or a fraction.

4. Donna read 90 pages of a novel in 120 minutes. At this rate, how many pages can Donna read during the next 80 minutes?

STEP 1 Find the rate in *pages per minute* that Donna reads.

STEP 2 Multiply the rate by 80, the number of minutes Donna will read.

5. Blake bought 4 pounds of hamburger for $5.40. At this rate, how much will Blake pay for 13 pounds?

STEP 1 **STEP 2**

$ per pound = ____________

6. Julia measured her heart rate. She counted 16 beats in 12 seconds. At this rate, how many times does Julia's heart beat each minute (60 seconds)?

STEP 1 **STEP 2**

beats per second = ____________

7. A 1.5-ounce piece of peanut butter candy contains 291 calories. About how many calories would a 5-ounce piece contain?

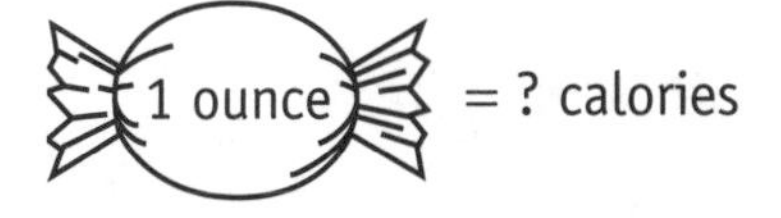

STEP 1 **STEP 2**

calories per ounce = ____________

Understanding Proportions

A **proportion** is made up of two equal ratios. If you add 2 pints of blue paint to 6 pints of white paint, the ratio of blue paint to white paint is $\frac{2}{6}$. You make the same color by adding 4 pints of blue paint to 12 pints of white paint. The colors are the same because the ratios of blue to white are equal fractions: $\frac{2}{6} = \frac{4}{12}$.

A proportion is usually written as a pair of equivalent (equal) fractions.

You read a proportion as two equal ratios connected by the word *as*.

$\frac{2}{6} = \frac{4}{12}$ is read "2 is to 6 *as* 4 is to 12."

In a proportion, the cross products are equal. To find the cross products, multiply each numerator by the opposite denominator.

Cross Multiplication	**Equal Cross Products**
$\frac{2}{6} \times \frac{4}{12}$	$2(12) = 6(4)$ $24 = 24$

Finding a Missing Number in a Proportion

To complete a proportion, write the letter n (or some other letter) for the missing number.

- Write the cross products. Note, one cross product contains the letter n.
- To find n, divide the complete cross product by the number that multiplies n.

EXAMPLE 1 Find the missing number. $\frac{n}{4} = \frac{9}{12}$

STEP 1 Cross multiply. $12n = 36\ (4 \times 9)$

STEP 2 To find n, divide 36 by 12.

$$n = \frac{36}{12} = 3$$

ANSWER: $\frac{\mathbf{3}}{4} = \frac{9}{12}$

(**Note:** In Step 1, $12n$ means $12 \times n$.)

EXAMPLE 2 Find the missing number. $\frac{2}{x} = \frac{6}{15}$

STEP 1 Cross multiply. $6x = 30\ (2 \times 15)$

STEP 2 To find x, divide to solve for x.

$$x = \frac{30}{6} = 5$$

ANSWER: $\frac{2}{\mathbf{5}} = \frac{6}{15}$

(**Note:** In Step 1, $6x$ means $6 \times x$.)

Cross multiply to see if each pair of fractions forms a proportion. Remember, in a proportion, cross products must be equal.

1. $\frac{3}{4} \stackrel{?}{=} \frac{8}{12}$ Yes No $\frac{2}{3} \stackrel{?}{=} \frac{4}{6}$ Yes No $\frac{2}{4} \stackrel{?}{=} \frac{3}{7}$ Yes No

2. $\frac{4}{5} \stackrel{?}{=} \frac{8}{10}$ Yes No $\frac{10}{16} \stackrel{?}{=} \frac{7}{8}$ Yes No $\frac{3}{7} \stackrel{?}{=} \frac{9}{21}$ Yes No

Write cross products for each proportion. The first one is done as an example.

3. $\frac{n}{2} = \frac{8}{16}$ $\frac{x}{3} = \frac{10}{15}$ $\frac{3}{n} = \frac{2}{8}$ $\frac{7}{10} = \frac{x}{5}$

$16n = 2 \times 8$
$16n = 16$

4. $\frac{8}{x} = \frac{3}{9}$ $\frac{n}{4} = \frac{48}{64}$ $\frac{7}{3} = \frac{x}{6}$ $\frac{n}{9} = \frac{26}{18}$

In each proportion below,

- **write the cross products**
- **divide to find the value of *n* or *x***

5. $\frac{n}{2} = \frac{3}{6}$ $\frac{x}{4} = \frac{9}{12}$ $\frac{5}{n} = \frac{20}{8}$ $\frac{12}{18} = \frac{x}{3}$

$n =$ ______ $x =$ ______ $n =$ ______ $x =$ ______

6. $\frac{3}{n} = \frac{9}{6}$ $\frac{4}{7} = \frac{x}{21}$ $\frac{n}{10} = \frac{3}{5}$ $\frac{x}{4} = \frac{20}{16}$

$n =$ ______ $x =$ ______ $n =$ ______ $x =$ ______

Using Proportions to Solve Word Problems

Proportions can be used to solve word problems involving comparisons. Remember to write the terms of each ratio in the correct order.

EXAMPLE 1 A punch recipe calls for mixing 3 cups of lemon-lime soda with 5 cups of punch. How much soda should Jamie mix with 18 cups of punch?

STEP 1 Write a proportion, where each ratio is $\frac{\text{cups of soda}}{\text{cups of punch}}$.

Let n stand for the unknown cups of soda.

$\frac{n}{18} = \frac{3}{5}$

STEP 2 Write cross products.

$5n = 18 \times 3 = 54$

STEP 3 Divide by 5 to solve for n.

$n = \frac{54}{5} = 10\frac{4}{5}$

ANSWER: $n = 10\frac{4}{5}$ **cups of soda**

EXAMPLE 2 Aaron earned $57 working 6 hours. At this same pay rate, how much would Aaron earn working 15 hours?

STEP 1 Write a proportion, where each ratio is $\frac{\text{amount paid}}{\text{hours worked}}$.

Let n stand for the unknown earnings.

$\frac{n}{15} = \frac{\$57}{6}$

STEP 2 Write cross products.

$6n = \$57 \times 15 = \$855$

STEP 3 Divide by 6 to solve for n.

$n = \frac{\$855}{6} = \$142.50$

ANSWER: $n = \$142.50$

Choose the proportion that can be used to solve each problem.

1. Pamela makes punch by mixing 2 cups of pineapple juice with 7 cups of orange juice. How many cups of pineapple juice (n) should Pamela mix with 16 cups of orange juice?

   a. $\frac{n}{2} = \frac{7}{16}$ b. $\frac{n}{7} = \frac{2}{16}$ c. $\frac{n}{16} = \frac{2}{7}$

2. On Monday Matsu drove 370 miles in 7 hours. At this rate, how many miles (m) can Matsu drive in 5 hours on Tuesday?

   a. $\frac{m}{5} = \frac{370}{7}$ b. $\frac{m}{7} = \frac{370}{5}$ c. $\frac{m}{370} = \frac{7}{5}$

Solve each problem.

7. Attendance at Basha's first four softball games was 257, 290, 312, and 305. What was the average attendance at these four games?

8. In the first game, Kelly got a bowling score that was 9 points below her average. In the second game, Kelly bowled a 126. What more do you need to know to be able to find Kelly's score in game 1?

9. In Blake's auto painting class, there are 26 men and 8 women.

   a. What is the ratio of *men to women?*

   b. What is the ratio of *women to men?*

10. Allison's cookie jar contains both chocolate chip and sugar cookies. Of the 22 cookies now in the jar, 14 are chocolate chip. Knowing this, determine the ratio of sugar cookies to chocolate chip cookies now in the jar.

11. Ella rode her bike a total of 42 miles in 3 hours.

    a. At what rate (in miles per hour) did Ella ride?

    b. Riding at this same rate, what distance can Ella ride in 5 hours?

12. Five out of seven students at Mona Lisa Art Academy are women. A total of 483 students attend this art school.

    a. Write a proportion that can be used to find the number of women (w) who attend Mona Lisa Art Academy.

    b. Solve the proportion for w.

PROBLEM-SOLVING STRATEGIES

Drawing a Picture

The first step in solving a problem is often to decide upon a strategy—a way of approaching or looking at the problem. In this chapter, several strategies will be discussed that you may find helpful.

There is a saying: "A picture is worth a thousand words." Drawing a picture is one of the most important math strategies. Think how much more difficult the example below would be without a drawing.

EXAMPLE Andre is driving his remote-control car. First, he drives the car 100 feet north of the starting point. He then drives the car 80 feet east. He then turns and drives the car 125 feet south. Next, he turns and drives the car 50 feet east. At this point, where is the car from the starting point?

? feet south ? feet east

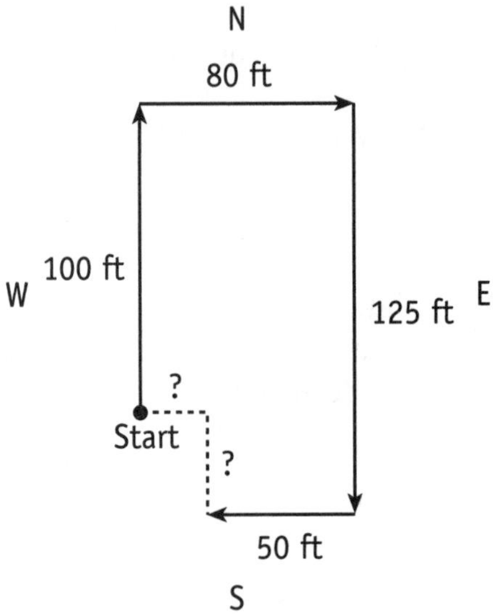

With all distances placed on the drawing, you can easily see where the car is from the starting point.

STEP 1 To determine how far the car is south of the starting point, subtract. 125 − 100 = **25 feet south**

STEP 2 To determine how far the car is east of the starting point, subtract. 80 − 50 = **30 feet east**

Use a drawing to solve each problem.

1. Amanda, Lauren, and Shauna all live on King Street, east of the library. Amanda lives 2.5 miles farther from the library than Lauren. Lauren lives 0.8 miles closer to the library than Shauna. Shauna lives 1.9 miles from the library. How far does Amanda live from the library?

Complete This Picture

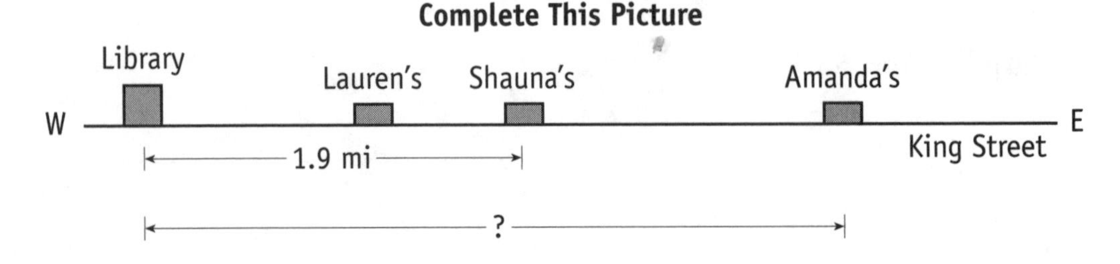

2. Henry knows a secret and tells it to Gary. Gary then tells the secret to three friends. Each of Gary's friends tells the secret to two friends. How many people, not counting Henry, know the secret?

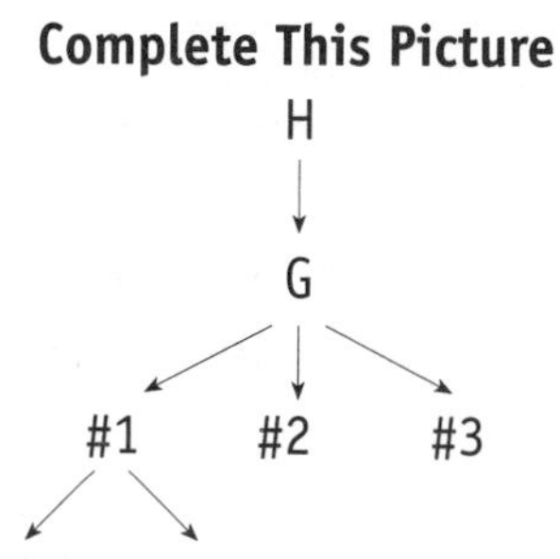

3. The distance around the figure (perimeter) at the right is 10 feet. What is the *least* number of squares you can add to the figure to make a new figure with a perimeter of 16 feet? Added squares must adjoin the figure along one or more whole sides. Make drawings on scratch paper to get your answer. Add to the drawing at the right to show your answer.

Add to This Picture

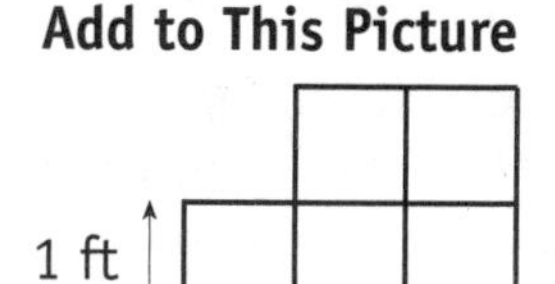

4. An airplane is flying at an altitude of 33,000 feet above sea level. A cloud is passing over the top of Mount Adams, the summit of which is 12,800 feet above sea level. If the cloud is 1,500 feet above the summit of Mount Adams, how high above the cloud is the airplane?

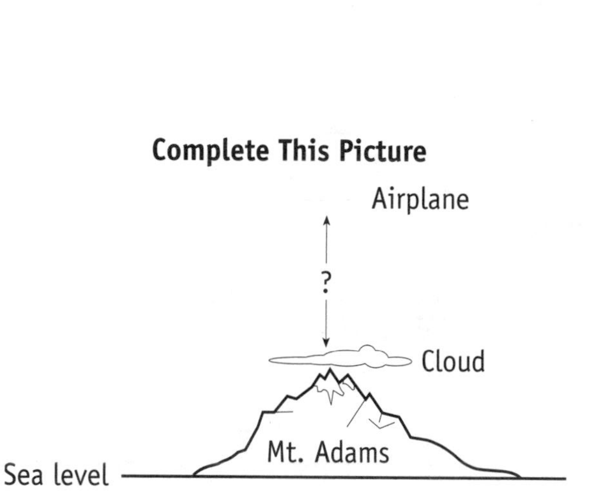

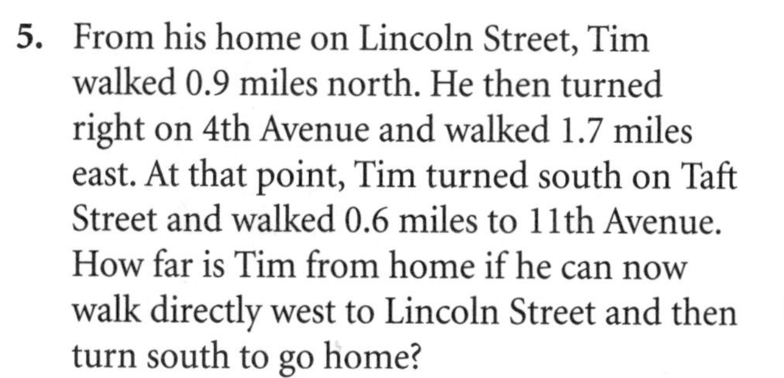
5. From his home on Lincoln Street, Tim walked 0.9 miles north. He then turned right on 4th Avenue and walked 1.7 miles east. At that point, Tim turned south on Taft Street and walked 0.6 miles to 11th Avenue. How far is Tim from home if he can now walk directly west to Lincoln Street and then turn south to go home?

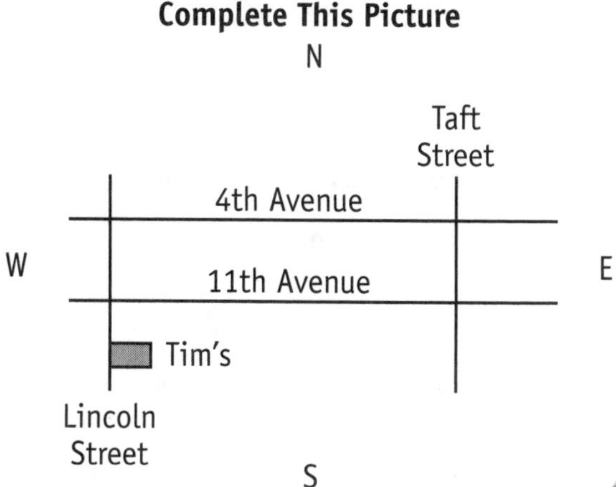

Making a List

Making a list is often the best way to find certain types of information. In math problems, a list can be used to help determine all possible combinations of numbers or seating arrangements, and so on. The example shows how a list is useful for this task.

EXAMPLE Steven knows that the three numbers of his lock combination are 7, 15, and 23. However, he can't remember the correct order of these numbers. How many different locker combinations are possible?

To find the number of different locker combinations, make a list of each possible combination. Make sure the list does not include the same combination twice.

First Number	Second Number	Third Number
7	15	23
7	23	15
15	7	23
15	23	7
23	7	15
23	15	7

Looking at this complete list, you count six possible locker combinations.

ANSWER: 6

Solve each problem by completing the list that's started for you.

1. David, Lennie, and Todd are going to a play. They have reserved-seat tickets D7, D8, and D9. How many seating arrangements are possible?

   Complete this list of seating arrangements.

D7	D8	D9
David	Lennie	Todd

2. Four students, Dolores, Teri, Keisha, and Andrea, are going to play in a checkers tournament. Each student will play each other student one game. How many games will be played in all?

   Complete the list of all possible games. (A game in which Dolores plays Teri is the same as the game in which Teri plays Dolores.)

   Games

   Dolores vs. Teri

3. For dinner Brenda has a choice of two drinks: milk or juice. She can choose one of two main dishes: chicken or spaghetti. And she can choose cake or ice cream for dessert. How many dinner combinations are available to Brenda?

   Complete the list of all possible dinner combinations.

Drink	**Main Dish**	**Dessert**
milk	chicken	cake

4. A penny has a heads-up side and a tails-up side. Suppose you toss three pennies into the air, and they land on the ground. How many ways can the three pennies land, assuming none land on an edge?

   Complete the list of all possible combinations of flips.

Penny 1	**Penny 2**	**Penny 3**
heads	heads	heads

Making a Table

A table is especially useful for organizing results of calculations. As the example shows, a table can be used to help you choose from among a list of possible answer choices.

EXAMPLE Francine sells gift packages of fruit. She sells a small package weighing 4 pounds and a large package weighing 7 pounds. Francine has 30 pounds of fruit left, and she wants to make a total of six more gift packages. How many of each size package can Francine make if she wants to use all 30 pounds of fruit?

To find out how many of each size package Francine needs, make a table. On the table, show each combination of packages that Francine could make. The table is shown at the right.

Small	Large	Total Pounds
0	6	$(0 \times 4) + (6 \times 7) = 42$
1	5	$(1 \times 4) + (5 \times 7) = 39$
2	4	$(2 \times 4) + (4 \times 7) = 36$
3	3	$(3 \times 4) + (3 \times 7) = 33$
4	2	$(4 \times 4) + (2 \times 7) = 30$
5	1	$(5 \times 4) + (1 \times 7) = 27$
6	0	$(6 \times 4) + (0 \times 7) = 24$

Looking at the completed table, you see that the answer is four small packages and two large packages.

ANSWER: 4 small packages, 2 large packages

Solve each problem by completing the table that's started for you.

1. a. Zander is buying a compact disc for $16.89. Zander pays for the disc with a $20 bill. Fill in the table to show five combinations of bills and coins Zander could receive in change.

$1.00 bill	25¢	10¢	5¢	1¢

b. Suppose Zander does not receive any $1 bills in change. What is the *least* number of coins Zander could receive in change?

2. Sherri spent $24.00 on seven posters. The small posters cost $3.00 each. The large posters cost $4.50 each. How many of each size poster did Sherri buy?

Small	Large	Total Price
1	6	$(1 \times 3) + (6 \times 4.5) = \$30.00$
2	5	$(2 \times 3) + (5 \times 4.5) =$
3	4	
4	3	
5	2	
6	1	

3. Jason took a quiz with two types of questions. They were worth either 6 or 9 points. Jason got 8 questions correct and scored 63 points. How many 9-point questions did Jason get correct?

6-Point	9-Point	Total Points
1	7	$(1 \times 6) + (7 \times 9) = 69$
2	6	$(2 \times 6) + (6 \times 9) =$
3	5	
4	4	
5	3	
6	2	
7	1	

4. The children's playroom has the same number of 3-legged stools as 4-legged stools. Mandy counted a total of 35 legs in all. How many of each type of stool is there?

3 Legs	4 Legs	Total Number of Legs
1	1	$(1 \times 3) + (1 \times 4) = 7$
2	2	$(2 \times 3) + (2 \times 4) =$
3	3	

Using Guess and Check

Sometimes you may not be sure how to solve a problem. One way is to guess the answer and then check your guess. Guessing and checking can often lead quickly to the correct answer.

EXAMPLE On Thursday evening Stacey called three friends on the phone. She remembers that the second call lasted 6 minutes longer than the first, and that the third call lasted 6 minutes longer than the second. If all three calls lasted a total of 66 minutes, how long was each call?

STEP 1 Guess that the first call lasted 15 minutes. Then the second call lasted 21 minutes (15 + 6) and the third call lasted 27 minutes (21 + 6). The sum of the three calls is 63 (15 + 21 + 27). This sum is less than 66. For your second try, guess a larger number.

STEP 2 Guess that the first call lasted 16 minutes. The second call is then 22 minutes (16 + 6) and the third call is 28 minutes (22 + 6). The sum is 66 (16 + 22 + 28) and is correct.

For the first guess, choose a number that seems about right. For the second guess, choose a larger or smaller number. Keep guessing until you get the correct answer.

ANSWER: first call: 16 minutes second call: 22 minutes third call: 28 minutes

Use guess and check to solve each problem.

1. Jenny has seven coins which have a total value of 39¢. What coins does Jenny have?

2. At a garage sale, Robbie received $1.50 in change after paying for three items with a $20 bill. Which items did Robbie buy?

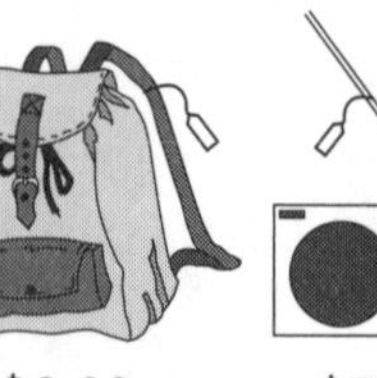

$5.00 $6.50 $6.00 $7.50

3. Wanda plans to spend part of her $78 on clothes and put part in savings. If she saves twice as much as she spends, how much will Wanda save?

4. Which two positive whole numbers, whose sum is 17, have the largest product?

5. Fred wants you to guess his address on Walnut Drive. He gives you three clues.

- The house number has three different digits.
- The digits total 21 when added.
- Each digit is 1 less than the digit before it.

Use these clues to determine Fred's street address.

6. Lois won the school-board election by receiving 450 more votes than Harmon. A total of 3,500 votes were cast. How many votes did each candidate receive?

Harmon: ______ Lois: ______

7. Amanda has 30¢ more than Stacey. Lauren has 15¢ more than Amanda. If together the three girls have $3.30, how much does each girl have?

Amanda: ______ Stacey: ______ Lauren: ______

8. Five members of the Zarate family paid a total of $14.50 for admission to the Jefferson Aquatic Center. How many of each type of ticket did the Zarate family buy?

Children: ______ Students: ______ Adults: ______

Jefferson Aquatic Center

Admission:

Adults	$4.50
Students	$2.50
Children (under 6)	$1.25

Working Backward

Some problems are most easily solved by starting with an end result and then working backward.

EXAMPLE For Jocelyn's party, Kari baked several trays of cookies. She kept 20 cookies at home and took the rest to the party. During the party, students ate half of the cookies. Students then took home 30 cookies and Kari took home the 12 cookies that remained. How many cookies did Kari bake?

STEP 1 Start with the number of cookies left over. To this number, add the number of cookies taken home (30).

12

$12 + 30 = 42$

You can conclude that 42 cookies remained after students ate half that were brought.

STEP 2 To find the number of cookies that were brought to the party, multiply 42 by 2. Kari brought 84 cookies.

$42 \times 2 = 84$

STEP 3 To find out how many cookies Kari baked, add 20 to 84.

$84 + 20 = 104$

ANSWER: Kari baked 104 cookies.

Solve each problem by working backward.

1. James is driving from Eugene to Seattle. During the trip, James plans to stop 1 hour for dinner and make a 15-minute stop for gasoline. If he didn't stop at all, he could make the drive in 5 hours. What time should James leave Eugene if he wants to be in Seattle at 9:00 P.M.?

2. The following results were obtained in a class poll.

   - The most popular dessert is ice cream.
   - Five more students prefer ice cream than cake.
   - Six more students prefer cake than cookies.

   Knowing that a total of three students prefer cookies, determine how many students are in the class.

3. As part of a fundraiser, four students took part in a lap swim. Kelly swam half as many laps as Aaron. Aaron swam 11 more laps than Susie. Susie swam 3 fewer laps than Mark. Mark swam a total of 36 laps.

   a. How many laps did Susie swim?

   b. How many laps did Kelly swim?

4. Diane brought a bag of stickers to school for a group project. Patti took six stickers. Lauren took twice as many stickers as Patti. Amanda took half of the stickers that were left, leaving only eight stickers in the bag.

   a. How many stickers did Amanda take?

   b. How many stickers did Diane bring to school?

5. A group of students went to the amusement park. At the park, one half of the students got a snack. Three of the remaining students went to the water slide. Of the students left, one third went to the fun house. The remaining four students rode the roller coaster.

   a. How many students went to the fun house?

   b. How many students got a snack?

   c. How many students went to the park?

Using a Number Pattern

For some problems, you must discover a pattern in a set of numbers. You can then determine other numbers in the set that are part of the same pattern.

EXAMPLE Posters are on sale at Paper Plus. Sale prices are shown in the table at the right. If this same price pattern continues, what will be the price per poster for a purchase of five posters?

Number of Posters	Price per Poster
1	$3.50
2	$3.25
3	$3.00
4	
5	

STEP 1 Notice that the price per poster decreases by $0.25 for each additional poster.

STEP 2 Subtracting $0.25, you find the price per poster of four posters to be $2.75. Subtracting $0.25 again, you find the price per poster of five posters to be $2.50.

ANSWER: $2.50 per poster for a purchase of five posters

For problems 1–4, complete the number pattern.

1. Write the next two numbers in the pattern below.

3	6	9	12	____	____
1st	2nd	3rd	4th	5th	6th

2. Write the 6th and 7th numbers in the pattern below.

1	2	4	7	11	____	____
1st	2nd	3rd	4th	5th	6th	7th

3. Write the 5th and 6th amounts in the pattern below.

$10	$19	$27	$34	____	____
1st	2nd	3rd	4th	5th	6th

4. Write the 5th and 6th pairs of numbers in the pattern below.

(1, 1)	(2, 4)	(3, 9)	(4, 16)	____	____
1st	2nd	3rd	4th	5th	6th

For problems 5–8, write your answers in the table.

5. The table shows the temperatures in San Francisco on a winter day. If this same pattern continues, what will be the temperature at 12 Noon and at 1:00 P.M.?

Time	Temperature
9:00 A.M.	44°F
10:00 A.M.	47°F
11:00 A.M.	50°F
12 Noon	
1:00 P.M.	

6. Ticket prices for the Hoover School Barbecue are shown at the right. If the price pattern continues, what would be the total cost of inviting four guests? (**Hint:** The price per guest decreases as the number of guests increases.)

Number	Total
1 guest	$5.00
2 guests	$9.00
3 guests	$12.00
4 guests	

7. A computer game awards a different point total for each level of difficulty. The point totals for levels A, B, C, and D are shown in the table. Assuming the same scoring pattern continues, what point totals are awarded at levels E and F?

Level Completed	Points Awarded
A	100
B	190
C	270
D	340
E	
F	

8. A frog is standing 256 centimeters from a pond. Each minute, the frog jumps half the distance remaining between itself and the pond. Complete the table to show the total distance the frog has jumped after the 3rd, 4th, and 5th minutes.

Minute that has just passed	1	2	3	4	5
Distance (cm) frog jumps during this minute	128	64			
Total distance (cm) frog has jumped by end of minute	128	192			

(**Hint:** At the start of the 3rd minute, how far is the frog from the pond?)

Using Logic

Logic problems challenge your thinking skills. To solve a logic problem, write things down, draw pictures, guess and check, use a table, look for patterns, and so on. Each problem is different. That's what makes them interesting.

EXAMPLE Four playing cards—a 7 of hearts, an 8 of hearts, a 9 of hearts, and a 10 of hearts—are lying face down. Use these clues to determine the order of the cards.

- The numbers on cards B and D are divisible by 2.
- The number on card B is less than the number on card A.

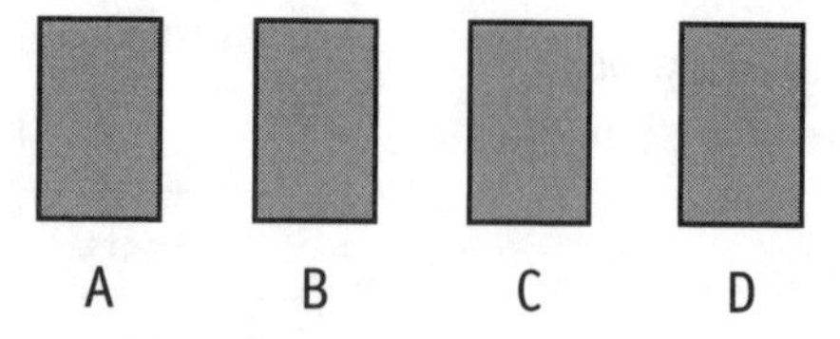

STEP 1 The first clue tells you that the number on card B is either the 8 or the 10.

STEP 2 The second clue tells you that the number on card A must be the 9 and that the number on card B must be the 8.

STEP 3 You can conclude that the number on card C is 7, and the number on card D is 10.

ANSWER: Card A: 9 Card B: 8 Card C: 7 Card D: 10

Use logic to solve each problem.

1. Caren, Lynda, and Sammy all work part-time at the Clothes Factory. During the 30 days in June,

   - Caren works June 2 and every other day after June 2.
   - Lynda works June 3 and every third day after June 3.
   - Sammy works June 4 and every fourth day after June 4.

   On how many June days will Caren, Lynda, and Sammy work on the same day?

2. Wanda has three quarters, two dimes, and two nickels. She has enough money to buy five stickers but not six. If the stickers are the same price, what is the least price the stickers could be?

3. Five teams of grade-school students competed in a swimming competition. Each group of students chose a team name. The results are as follows.

   - The Water Fleas scored more points than the Friendly Fish.
   - The Dolphins scored enough to take second place.
   - The Whale Watchers scored more points than the Swift Salmons.
   - The Friendly Fish finished ahead of the Whale Watchers.

   Use these clues to determine the finish order of the five teams.

   1st ____________ 2nd ____________ 3rd ____________

   4th ____________ 5th ____________

4. Jessica wants to neatly arrange her mystery novels collection into several equal stacks. Here is what happens when she tries to do this.

   - stacks of 2: one novel left over
   - stacks of 3: one novel left over
   - stacks of 5: equal stacks, no novels left over

   Jessica has more than 25 novels, but fewer than 75. Use the clues above to determine exactly how many novels Jessica has.

5. Raoul and Esther are going to paint a room. By himself Raoul can finish the room in 3 hours. By herself Esther can finish the room in 2 hours.

   a. Working alone for 1 hour, what fraction of the job could Raoul complete?

   b. Working alone for 1 hour, what fraction of the job could Esther complete?

   c. Working together for 1 hour, what fraction of the job could Raoul and Esther complete?

   d. Working together, about how long will it take Raoul and Esther to complete the job?

Problem-Solving Strategies Review

Work each problem and check your answers. Correct any errors.

1. Jody gave Tina a treasure map with the following instructions.

   To find the treasure, start at the large rock. From the rock, walk 40 feet north. Then walk 35 feet west. Next, walk 60 feet south. At that point, walk 20 feet east. Finally, walk 15 feet north and 15 feet east. The treasure is buried beneath your feet!

   Give the distance and direction from the rock of the point at where the treasure is buried.

   ______ feet ______ of the rock
   distance direction

Making a Drawing May Help

N

W large rock E

S

2. Todd remembers that the three numbers in Millie's telephone number area code are 1, 4, and 5. He can't remember what order the numbers are in. How many area code combinations are possible using these three numbers?

Writing a List May Help

3. A bus leaves the community center at 7:30 A.M. and every 45 minutes after. The ride to the downtown stop takes 20 minutes. If Lauren wants to ride from the community center and be downtown shortly before 10:15 A.M., what time should she catch the bus?

Making a Table May Help

Bus Leaves Community Center	Bus Arrives Downtown

4. Together, Anna and Jessie sold 462 boxes of Girl Scout cookies. If Jessie sold 18 more boxes than Anna, how many boxes did each girl sell?

Anna: ______ Jessie: ______

Guess and Check May Help

5 In 1980 rent at Alpine Apartments was only half what it was in 1990. In 1995 rent was $150 more per month than it was in 1990. Today's rent of $950 per month is $100 more per month than the rent in 1995. How much was rent at Alpine Apartments in 1980?

Working Backward May Help

6 Monique wants to rent four movies for the weekend from Movieland Video. Assuming the price pattern continues, what can Monique expect to pay?

Movieland Video Rentals

Number of Movies	Weekend Rate
1	$2.98
2	$4.77
3	$6.56

7. Four playing cards—the 3, 4, 6, and 8 of clubs—are lying face down. Use these clues to determine the order of the cards.

- The numbers on cards A and C are divisible by 3.
- The number on card D is less than the number on card C.

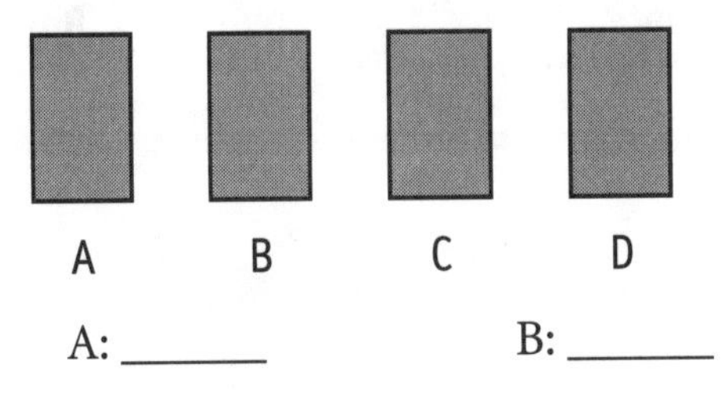

A: ______ B: ______ C: ______ D: ______

DATA ANALYSIS

Organizing Data on a List

Data can be names, numbers, or other information.

Data analysis, also called **statistics,** deals with organizing, interpreting, and using data. A group of data values, often called a **set,** may be organized in alphabetical (by letter) or numerical (by number) order.

EXAMPLE Five students in Amy's class sold candy as a fund-raising activity. Amy organized the results into two lists: a list *alphabetized by last names* and a list in *order of total sales.*

Fund-Raising Results

Ron Pastega	$57.25
Manuel Valdez	$75.00
Keisha Jones	$64.50
Mai-Lin Yu	$46.75
Kaitlin Brown	$49.00

Alphabetical List

Brown, Kaitlin	$49.00
Jones, Keisha	$64.50
Pastega, Ron	$57.25
Valdez, Manuel	$75.00
Yu, Mai-Lin	$46.75

Numerical List

$75.00	Manuel Valdez
$64.50	Keisha Jones
$57.25	Ron Pastega
$49.00	Kaitlin Brown
$46.75	Mai-Lin Yu

For each original list, make two new lists, organized as indicated.

1. Original List

Invoice #	Name
B376	Lucy White
B483	Basha Larnie
B294	Staci Lin
B349	Yoon Yu
B429	Stan Eachus

a. by invoice #, smallest first

Invoice #	Name

b. alphabetical by last name

Name	Invoice #

2. Original List

Balance	Company
$2,386	Rivera Co.
$1,965	Thompson's
$2,800	Harleton
$1,795	ELB, Inc.
$2,498	L-Plus, Ltd

a. by balance, largest first

Balance	Company

b. alphabetical by company name

Company	Balance

Using a Tally Sheet

A **tally sheet** is a record of slashes (called **tally marks**) that are used to keep track of a count. Tally marks are usually drawn in groups of five.

EXAMPLES

||| represents 3 𝍸 represents 5 (4 + 1) 𝍸 𝍸 |||| represents 14 (5 + 5 + 4)

Using a Tally Sheet

EXAMPLE Amanda asked 200 students how they get to school. She recorded each answer on a tally sheet. According to the poll, how many students ride the bus?

Transportation Poll of 200 Students

Bus	Car	Walk	Other
𝍸 𝍸 𝍸 𝍸 𝍸 𝍸 𝍸 𝍸 𝍸 𝍸 𝍸 𝍸 𝍸 𝍸 𝍸 𝍸	𝍸 𝍸 𝍸 𝍸 𝍸 𝍸 𝍸 𝍸 ||||	𝍸 𝍸 𝍸 𝍸 𝍸 𝍸 𝍸 𝍸 𝍸 𝍸 𝍸 |	𝍸 𝍸 𝍸 𝍸

According to the poll, **80 of the 200 students ride the bus.** (To count tally marks, count by 5s and then add any additional tallies that aren't in a group of 5.)

Use the tally sheet above to answer the following questions.

1. **Count Tally Marks** According to the poll, how many students answered

   Car?

   Walk?

   Other?

2. **Compare Totals** How many more students walked than rode in a car?

3. **Find a Ratio** What is the ratio of students answering Other to those answering Bus?

4. **Find a Percent** What *percent* of students answered

   Bus? Walk?

   Car? Other?

Finding Typical Values

Sometimes you may be interested in knowing only the most typical value of data. For example, look at the ticket prices shown at the right. What would you say is the most typical price for a ticket at these theaters?

In the study of math, there are three ways to give a typical value: **mean, median,** and **mode.**

Ticket Prices	
Broadway Cinema	$6.25
Cinema Center	$6.75
Hollywood Theater	$7.00
Movie Land	$5.75
Showtime Classics	$6.75

Mean

The most commonly used typical value is the **mean,** another word for **average.** (You first studied average on page 42.) To find the mean of a set (group) of numbers

- Add the numbers in the set.
- Divide the sum by the number of numbers in the set.

In most problems, the mean does not equal any number in the set. However, the mean is usually close to the middle number of the set.

EXAMPLE 1 To find the **mean** ticket price, follow these steps.

STEP 1 Add.

```
  $6.25
   6.75
   7.00
   5.75
+  6.75
 $32.50
```

STEP 2 Divide.

```
   $6.50  ← mean
5)$32.50  ← sum
↑
└─number in set
```

The mean ticket price is **$6.50.**

Median

The **median** is the middle value of a set of numbers.

- If a set contains an odd number of numbers, the median is the middle number.
- If a set contains an even number of numbers, the median is the average of the two middle numbers.

EXAMPLE 2 To find the **median** ticket price, follow these steps.

STEP 1 Arrange the prices in order, from least to greatest.

$5.75, $6.25, $6.75, $6.75, $7.00

↑ middle value (points to the first $6.75)

STEP 2 Because there is an odd number (5) of values, the median is the middle value.

The median ticket price is **$6.75.**

Mode

The **mode** of a set of numbers is the number that appears the most times. If no number appears more than once, a set has no mode.

EXAMPLE 3 The **mode** of the ticket prices is **$6.75,** the only amount appearing more than once.

In each set of data, find the mean, median, and mode (if there is one).

1.

Student Test Scores	
Laura	90
Colin	84
Stacey	87
Greg	81
Karin	83

Mean: ______

Median: ______

Mode: ______

2.

Elementary School Enrollment	
Washington Elementary	400
Lincoln Elementary	350
Mountain View Elementary	370
Hoover Elementary	400
Crescent Valley Elementary	390

Mean: ______

Median: ______

Mode: ______

Find the mean and median in each set of measurements below. In problems 4 and 5, the median is the average of the two middle values.

3.

Package Weights (1 lb = 16 oz)
4 pounds 10 ounces
6 pounds 12 ounces
5 pounds 4 ounces
8 pounds 14 ounces
7 pounds 5 ounces

Mean: ______

Median: ______

4.

Student Heights (1 ft = 12 in.)
5 feet 1 inch
4 feet 10 inches
5 feet 2 inches
4 feet 11 inches

Mean: ______

Median: ______

5.

Liquid Amounts (1 c = 8 fl oz)
2 cups 5 fluid ounces
3 cups 2 fluid ounces
1 cup 6 fluid ounces
2 cups 3 fluid ounces

Mean: ______

Median: ______

Lia kept a tally of the number of students who signed up for an art class given each month.

March	April	May	June
𝍸 𝍸 𝍸 III	𝍸 𝍸 𝍸 𝍸 𝍸 I	𝍸 𝍸 𝍸 𝍸 𝍸 III	𝍸 𝍸 𝍸 𝍸

6. How many students signed up each month?

March April

May June

7. What is the mean number of students taking the class each month?

8. What is the median number of students taking the class each month?

Reading a Table

A **table** displays data in labeled rows and columns.

A **row** is read from left to right.

Row $2.50 $3.75 $1.82 $2.65

Read across. →

A **column** is read from top to bottom.

Column

235 Read
184 down.
302 ↓
195

Look at the nutrition table below. The table contains a **title,** row and column **labels** (headings), and **numerical data.**

Nutritional Values of a 3-Ounce Serving of Selected Fish

Fish	Calories	Fat (g)	Cholesterol (mg)	Sodium (mg)
bass	82	2.0	68	59
halibut	119	2.5	35	59
mackerel	223	15.1	64	71
ocean perch	80	1.4	36	64
rainbow trout	100	2.9	48	23
salmon	99	2.9	44	57
swordfish	103	3.4	33	76
yellowfin tuna	92	0.8	38	31

g = grams
mg = milligrams

EXAMPLE 1 How many milligrams of sodium does a 3-ounce serving of mackerel contain?

Find the intersection of the row labeled *mackerel* and the column labeled *Sodium.*

Sodium (mg)
⋮ Read
59 down.
mackerel . . . 64 71
Read across.

ANSWER: 71 mg

EXAMPLE 2 Which of the listed fish is lowest in fat?

Scan down the column labeled *Fat* and choose the smallest number: 0.8. Read the label of the row that contains 0.8.

Fat (g)
⋮ Read
yellowfin tuna 92 0.8 down.
Look to the left for the row label.

ANSWER: yellowfin tuna

For problems 1–4, refer to the nutrition table on page 74.

1. **Read a Data Value** How many calories does a 3-*ounce* serving of salmon contain?

2. **Compare Data Values** Which of the listed fish is lowest in cholesterol?

3. **Estimate a Ratio** What is the *approximate* ratio of sodium in swordfish to sodium in rainbow trout?

   a. $\frac{2}{1}$ b. $\frac{3}{1}$ c. $\frac{4}{1}$

4. **Find an Average** What is the average number of calories in a 3-ounce serving of the fish listed in the table?

For problems 5–8, refer to the table below.

Calories Burned per Hour from Walking

Personal Weight (lb)	Walking Rate (miles per hour) 3.0	3.5	4.0	4.5	5.0
100	162	181	201	306	413
120	195	218	241	367	496
140	228	254	281	429	578
160	260	291	322	490	661
180	293	327	362	552	744
200	326	364	402	613	827

Did You Know . . . ?

- To *interpolate* is to guess a data value that lies between two given values.
- To *extrapolate* is to guess a data value that lies outside the range of given values.

5. **Compare Data Values** How many more calories does a 140-pound woman burn each hour walking 4 miles per hour rather than 3 miles per hour?

6. **Estimate a Ratio** Estimate the ratio of calories burned by a 200-pound man to those burned by a 100-pound man while walking?

7. **Interpolate** About how many calories are burned each hour by a 130-pound woman walking 3 miles per hour?

8. **Extrapolate** About how many calories are burned each hour by a 250-pound man walking 4 miles per hour?

Reading a Pictograph

A **pictograph** shows data as small pictures (**symbols**). The value of the symbol is given in a **key.** Half of a symbol stands for half of a whole symbol's value.

Pictographs are used to compare data. They do not give exact values because of rounding. Small differences in values are not shown.

The pictograph below has a **title,** row **labels,** and data **symbols.** Monthly sales are shown as rows of symbols. Because half symbols are used, monthly sales are rounded to the nearest 50 (half symbol). To find the value of a row, multiply the number of symbols by the value of the symbol as shown in the key.

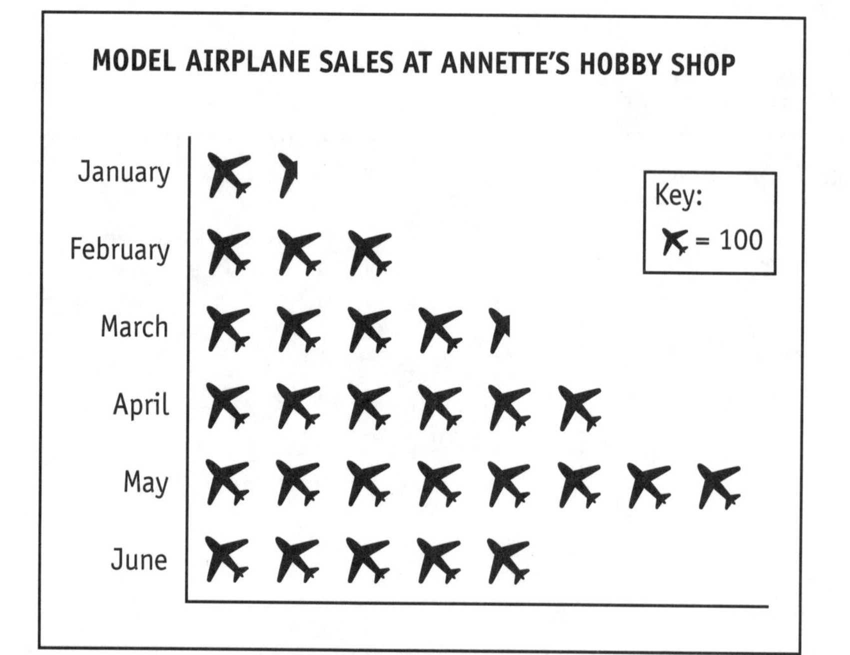

EXAMPLE 1 About how many model airplanes did Annette's Hobby Shop sell during March?

STEP 1 Count the number of symbols in the March row.

$4\frac{1}{2}$

STEP 2 Multiply the symbol value (100) times the number of symbols.

$100 \times 4\frac{1}{2} = 450$

ANSWER: about 450 model airplanes

EXAMPLE 2 What is the average number of airplanes sold per month during the 6 months shown?

STEP 1 Multiply the total number of symbols

$(27 + \frac{1}{2} + \frac{1}{2} = 28)$ by 100.

$28 \times 100 = 2{,}800$ (total number sold)

STEP 2 Divide 2,800 by 6.

$2{,}800 \div 6 \approx 467$

ANSWER: about 467 planes per month

For problems 1–4, refer to the pictograph on page 76.

1. **Compare Data Values** How many more model airplanes did Annette's Hobby Shop sell during May than during March?

2. **Estimate a Ratio** Estimate the ratio of model airplane sales during Annette's best month to sales during Annette's worst month.

3. **Find a Median** What is the median number of monthly sales of model airplanes at Annette's Hobby Shop?

4. **Analyze** What is the greatest number of model airplanes that Annette's actually may have sold during May? (**Remember:** The pictograph is based on rounded numbers.)

For problems 5–8, refer to the pictograph below.

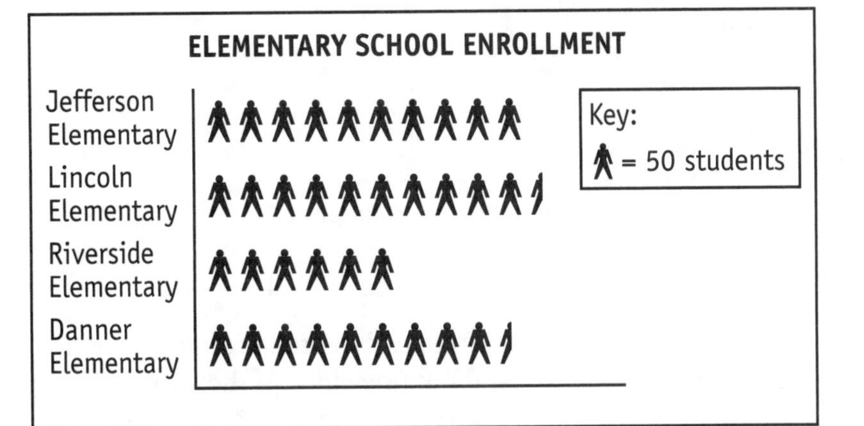

5. **Compare Data Values** About how many more students attend Lincoln Elementary than Riverside Elementary?

6. **Analyze** Franklin Elementary has 464 students. How many symbols would be used to represent Franklin Elementary?

7. **Find a Mean** What is the mean enrollment for the four elementary schools listed?

8. **Estimate a Percent** About what percent of all students in the four schools listed attend Riverside Elementary?

Reading a Bar Graph

A **bar graph** uses bars to display data for easy comparison. Data bars may be drawn vertically (up and down) or horizontally (across).

The bar graph below has a **title** and **labels** along each **axis**—the side of a graph along which words or numbers are listed. Data value is represented by the height of each bar.

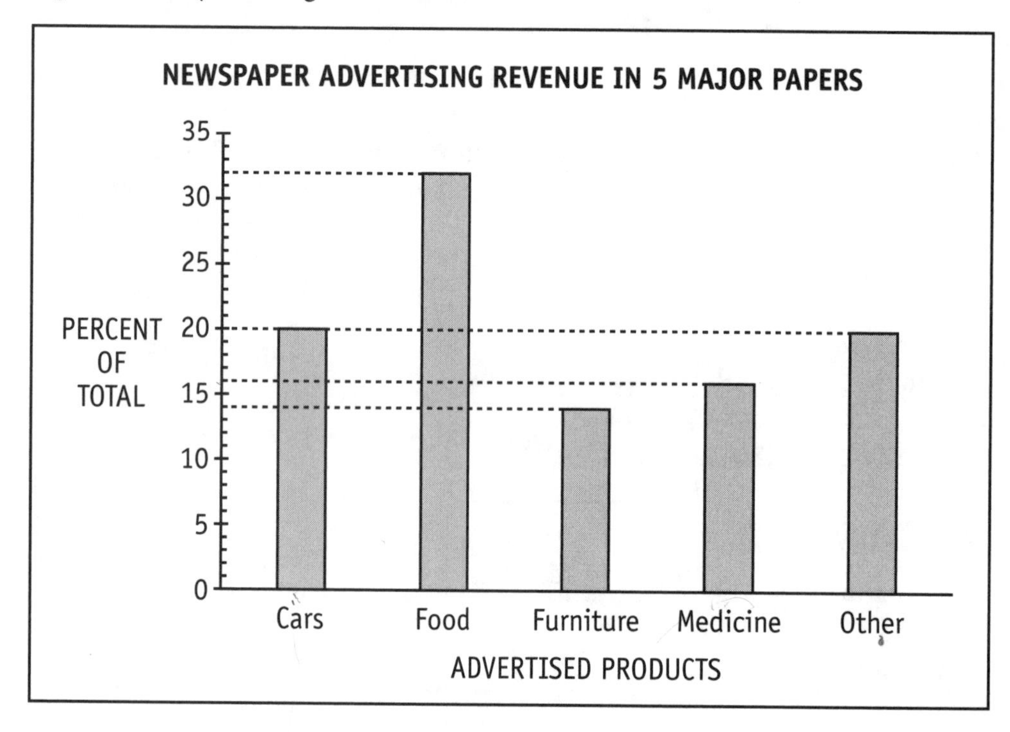

EXAMPLE 1 What percent of the total advertising revenue comes from the advertising of medicine?

STEP 1 Locate the bar above the word *Medicine.*

STEP 2 Read the height of the bar as a number on the vertical axis.

ANSWER: 16 percent

EXAMPLE 2 Which advertised product brings in about one-third of the total advertising revenue?

STEP 1 Write the fraction $\frac{1}{3}$ as a percent.

$\frac{1}{3} = 33\frac{1}{3}\%$

STEP 2 Choose the bar most nearly equal to 33%.

Food (32%)

ANSWER: Food

For problems 1–4, refer to the bar graph above.

1. **Compare Data Values** What is the difference in percent of advertising revenue received from cars than that received from medicine?

2. **Estimate a Ratio** Estimate the ratio of advertising revenue received from food to revenue received from furniture.

3. **Find Part of a Total** For each $500,000 of advertising revenue, how much is received from the advertising of medicine?

4. **Analyze** What advertised product brings in about one-sixth of the total advertising revenue?

For problems 5–8, refer to the bar graph below.

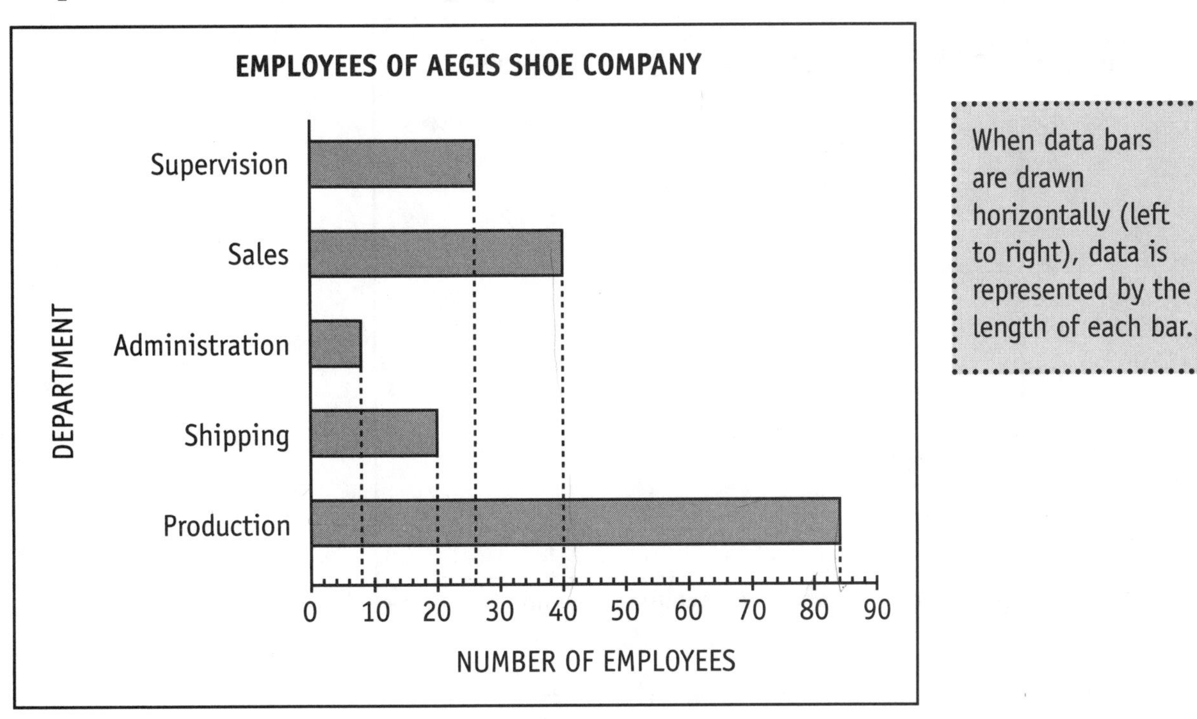

5. **Read a Data Value** How many Aegis employees work in production?

6. **Compare Data Values** How many more Aegis employees work in sales than work in administration?

7. **Estimate a Ratio** Estimate the ratio of Aegis employees working in supervision to those in production.

   a. $\frac{1}{5}$ **b.** $\frac{1}{4}$ **c.** $\frac{1}{3}$ **d.** $\frac{1}{2}$

8. **Estimate Percent** Approximately what percent of Aegis employees work in sales? (**Hint:** As your first step, find the total number of employees.)

Reading a Line Graph

A **line graph** displays data as points connected by short lines. The value of each point is read as a number along each axis. Line graphs often show how something changes over time.

The line graph below has a **title, labels** along each axis, and six **data points.**

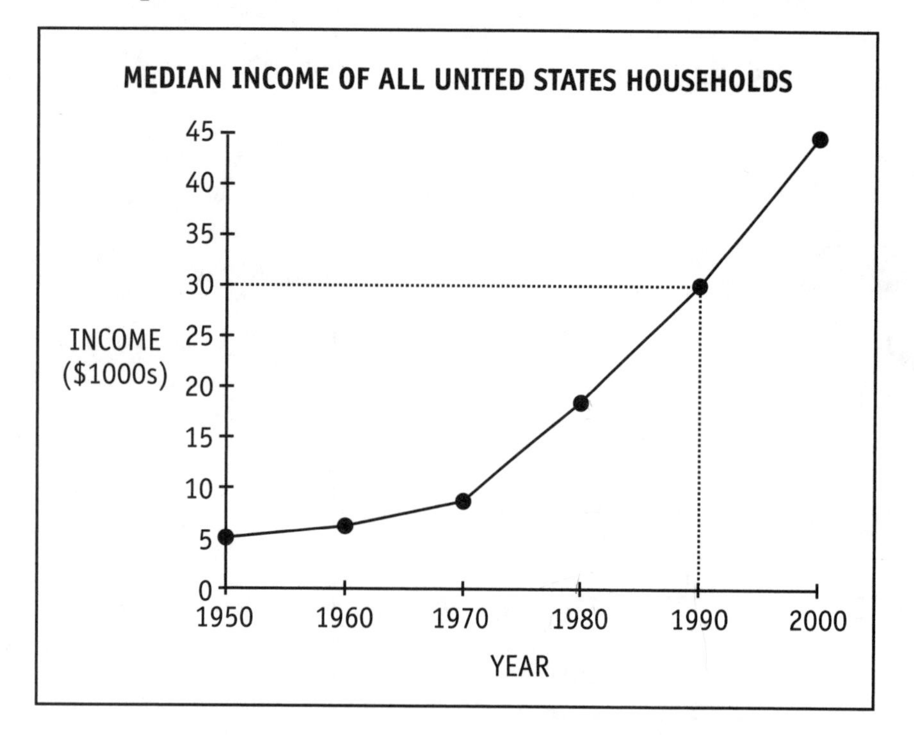

EXAMPLE 1 What is the approximate median income of all households in the United States in the year 1990?

STEP 1 Find the point on the line directly above the year 1990.

STEP 2 From the point on the line, read directly across to the value on the vertical axis.

ANSWER: about $30,000

EXAMPLE 2 Estimate the ratio of the median income in 1980 to that in 1950.

STEP 1 Find the median income for the years 1980 and 1950.

1980 ≈ $18,000   1950 ≈ $5,000

STEP 2 Form a ratio.

$$\frac{\$18{,}000}{\$5{,}000} = \frac{18}{5}$$

ANSWER: $\frac{18}{5}$

For problems 1–4, refer to the line graph above.

1. **Analyze** On the line graph above, you can read data points to the nearest _______.

   a. $100
   b. $1,000
   c. $10,000

2. **Compare Data Values** By about how much (in $1000s) did median household incomes increase between 1950 and 2000?

3. **Analyze** During which decade (10-year period) did median household incomes first rise above $25,000?

   a. 1960s **b.** 1970s **c.** 1980s

4. **Estimate Percent Increase** Estimate the *percent increase* in median household incomes that occurred between 1980 and 1990.

For problems 5–8, refer to the line graph below.

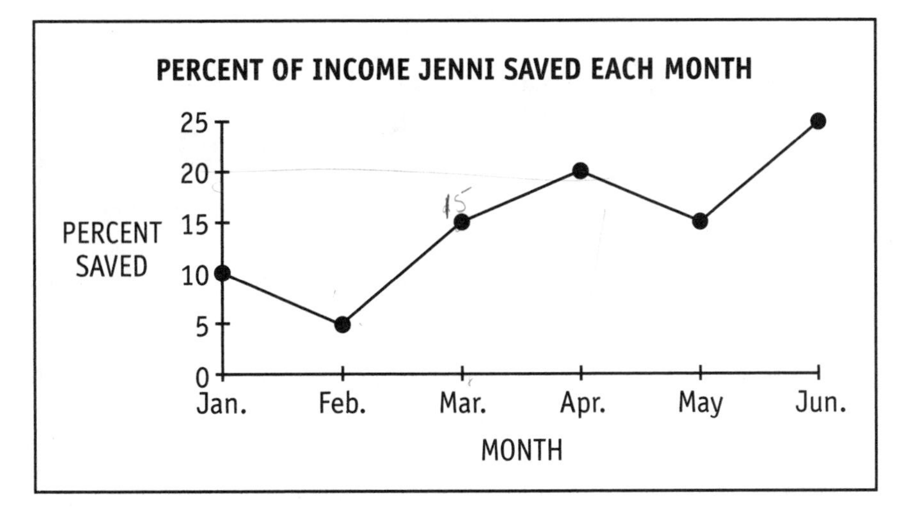

5. **Compare Data Values** What is the difference in percent at which Jenni saved in May compared to what she saved in February?

6. **Find an Average** What average percent of her monthly income did Jenni save during the 6 months shown?

7. **Find Part of a Total** During March Jenni earned $180. What amount (in dollars) did Jenni save during March?

8. **Analyze** During one of the months listed, Jenni saved $75 of the $375 she earned. Which month was this?

Reading a Circle Graph

A **circle graph** displays data as sections (parts) of a divided circle. A circle graph shows how a whole amount is made up of several parts.

Circle graphs usually show data in one of two ways.

- **percents,** where the whole equals 100%
- **cents per dollar,** where the whole equals $1.00

The circle graph below has a **title** and **labels** for each of the six sections, each label being a name and a percent.

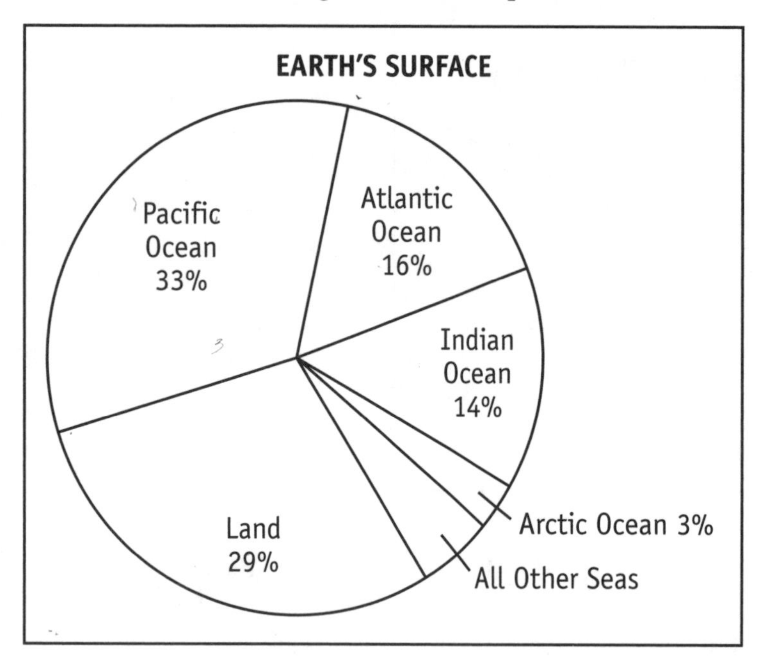

> **Remember:**
> The sections of a circle divided into percents must total 100%.

EXAMPLE 1 What percent of the earth's surface is covered by land?

Read the percent value in the section labeled *Land.*

ANSWER: 29%

EXAMPLE 2 What total percent of the earth's surface is covered by the three largest oceans?

Add the values of the Pacific, Atlantic, and Indian Oceans.

33% + 16% + 14% = 63%

ANSWER: 63%

For problems 1–4, refer to the circle graph above.

1. **Compare Data Values** Which ocean covers more of the surface of the earth than all land formations combined?

2. **Estimate a Ratio** What is the approximate ratio of the earth's surface covered by the Pacific Ocean to that covered by the Atlantic Ocean? 

3. **Analyze** What percent of the earth is covered by water represented by the section labeled All Other Seas?

4. **Find Part of the Total** If the earth has a surface area of about 200 million square miles, how many million square miles are covered with land?

For problems 5–8, refer to the circle graph below.

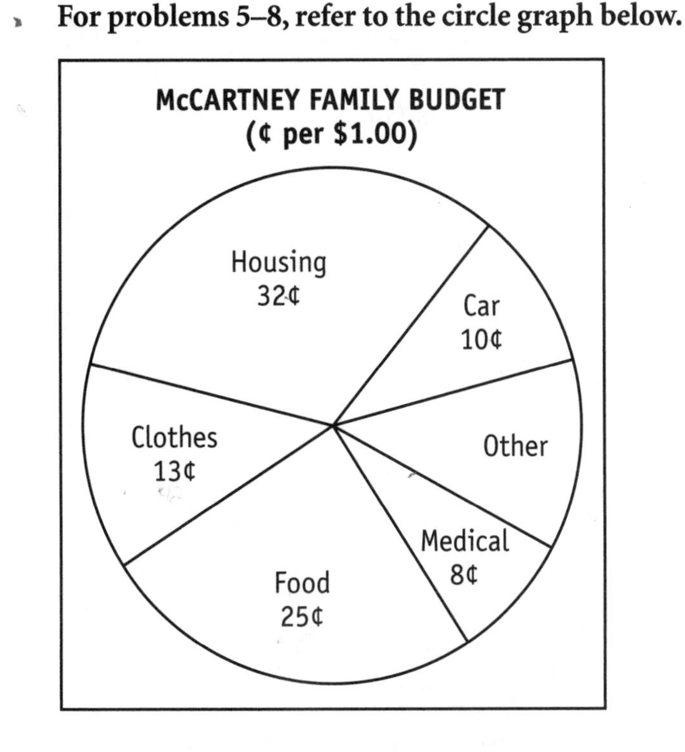

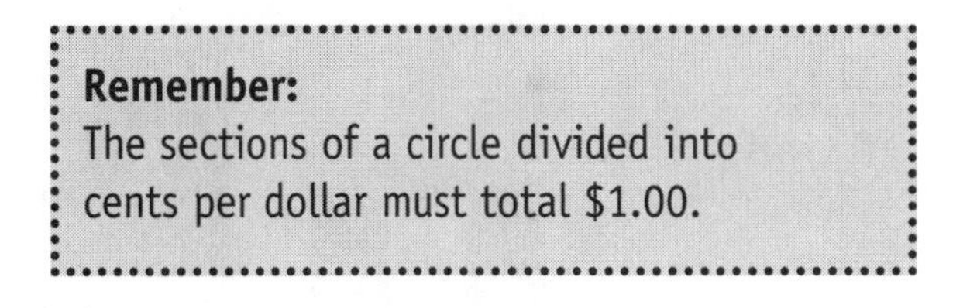

Remember:
The sections of a circle divided into cents per dollar must total $1.00.

5. **Compare Data Values** For each $1.00 the McCartneys spend, how much more do they spend on housing than on food?

6. **Find a Ratio** What is the ratio of the amount the McCartneys spend each month on food to the amount they spend on their car?

7. **Analyze** How many cents out of each dollar do the McCartneys spend in the category labeled *Other*?

8. **Find Part of a Total** If the McCartney monthly income is $1,650, how much do they spend each month on housing?

Changing Data from One Form to Another

Data can be shown in more than one way. For example, the bar graph and the circle graph below show the same information.

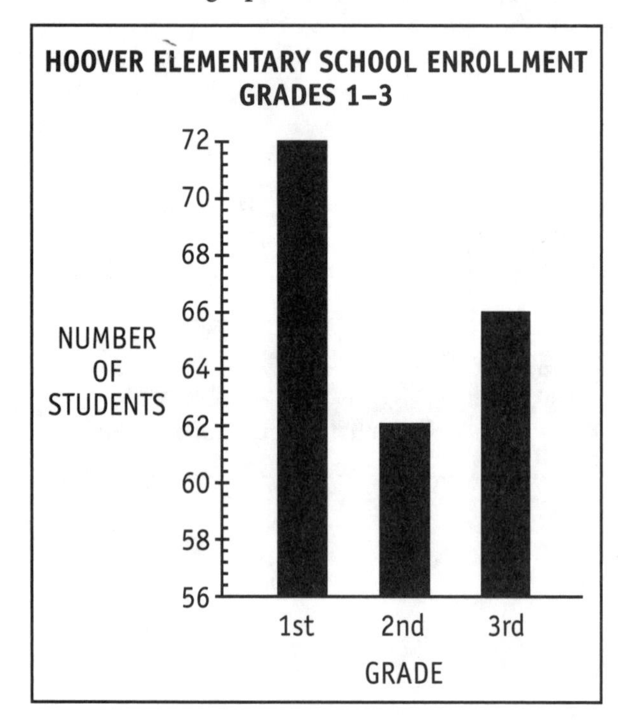

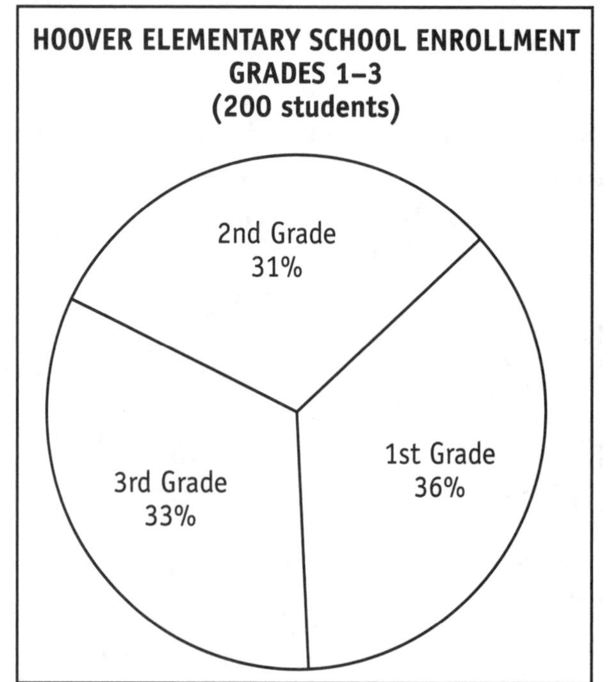

EXAMPLE 1 Use the bar graph to figure out what percent of the students are in the 1st grade.

STEP 1 Add to find the total number of students.

$72 + 62 + 66 = 200$

STEP 2 Change the fraction $\frac{72}{200}$ $\left(\frac{\text{1st grade}}{\text{total}}\right)$ to a percent. An easy way to do this is to multiply by 100%.

$\frac{72}{200} \times 100\% = \frac{72}{2}\% = 36\%$

ANSWER: 36%

EXAMPLE 2 Use the circle graph to figure out how many students are in the 2nd grade.

STEP 1 Change 31% to a fraction.

$31\% = \frac{31}{100}$

STEP 2 Multiply 200 (the total number of students) by $\frac{31}{100}$.

$200 \times \frac{31}{100} = 62$

ANSWER: 62

For problems 1 and 2, refer to the graphs above.

1. Use the bar graph to figure out what percent of the students are in 3rd grade. Check your answer on the circle graph.

2. Use the circle graph to figure out how many students are in the 3rd grade. Check your answer on the bar graph.

3. Use data from the pictograph to complete the line graph to show the number of books read each month by students in Keisha's class.

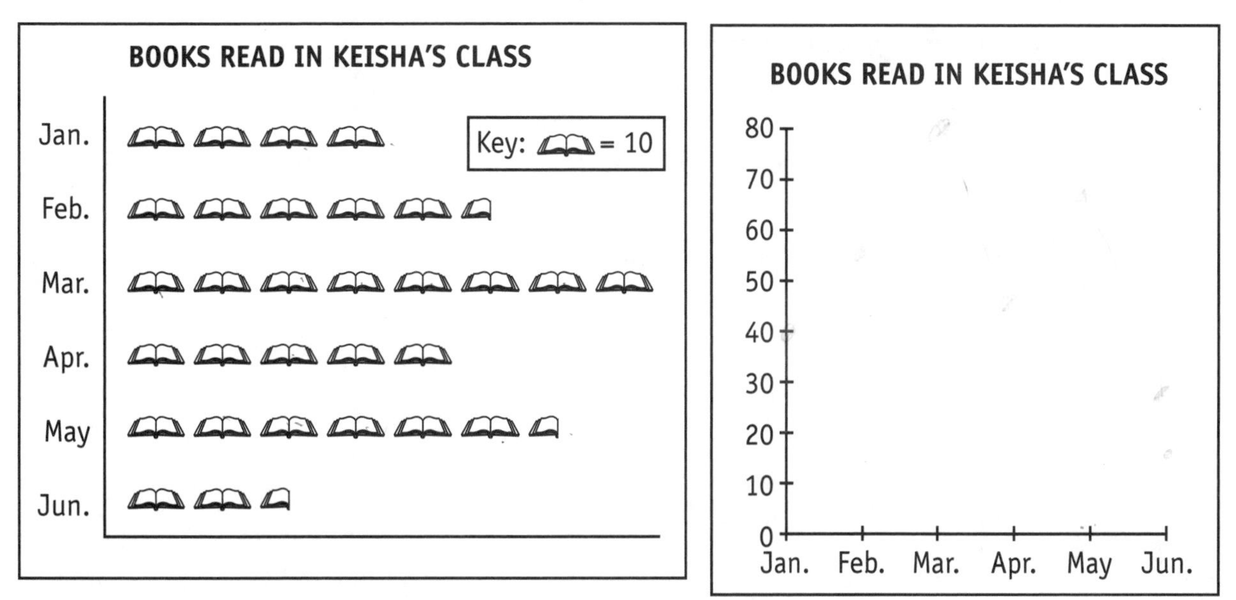

4. Use data from the bar graph to complete the circle graph to show the number of cents out of each dollar that was spent on each purchased item. Punch is done as an example.

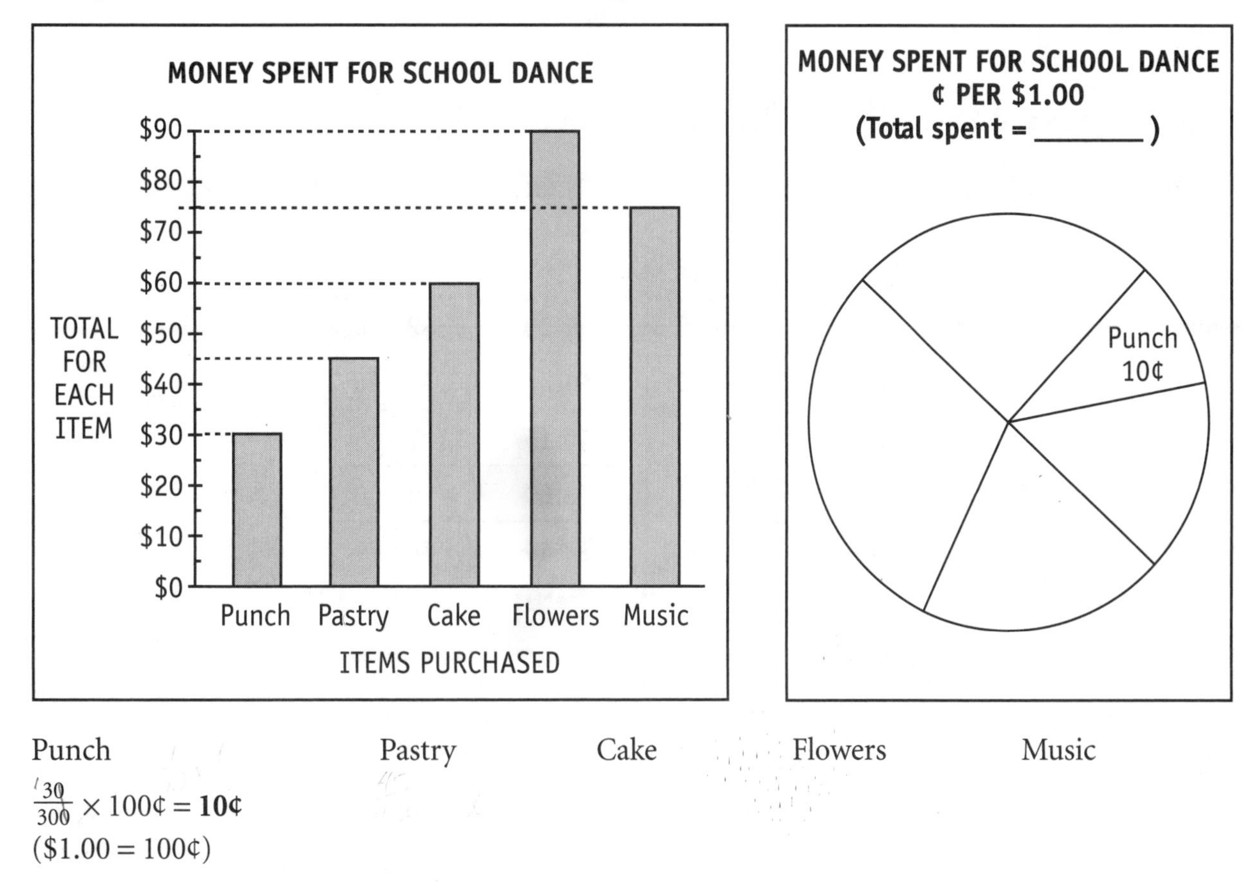

Punch | Pastry | Cake | Flowers | Music

$\frac{30}{300} \times 100¢ = \mathbf{10¢}$

($1.00 = 100¢)

Data Analysis Review

Work each problem and check your answers. Correct any errors.

1. Write the following list in order of *amount*, largest amount first.

   Caren, $939; Joclyn, $794; Kaitlin, $875; Susan, $1,049; Dee, $892

   Amount **Name**

2. Find the mean, median, and mode in the list of pizza prices shown at the right.

   Mean:

   Median:

   Mode:

Maria's Pizza	$14.00
Tino's Pizza	$12.50
Pizza Kitchen	$13.00
House of Italy	$15.00
Mona Lisa	$12.50

For problems 3 and 4, refer to the table below.

Item	Original Price	Discount Rate	Sale Price	Sales Tax (6%)	Total Price
Shirt	$25.00	20%	$20.00	$1.20	$21.20
Pants	$30.00	30%	$21.00	$1.26	$22.26
Socks	$5.00	50%	$2.50	$0.15	$2.65
Coat	$80.00	25%	$60.00	$3.60	$63.60
Shoes	$45.00	10%	$41.50	$2.49	$43.99

3. Including sales tax, what is the total cost of a pair of pants?

4. On which item is the actual dollar amount of savings the greatest?

For problems 5 and 6, refer to the pictograph at the right.

5. During the weeks shown, what is the average number of phone calls made per week?

6. What is the ratio of the calls made during Week 4 to the calls made during Week 5? (Express the ratio in lowest terms.)

PHONE CALLS MADE IN THOMPSON HOUSEHOLD DURING 5-WEEK PERIOD

Week 1
Week 2
Week 3
Week 4
Week 5

Key: = 6

For problems 7 and 8, refer to the bar graph at the right.

7. How much did the population of the United States increase between 1960 and 2000?

8. In which year do you think the population of the United States will reach 300 million?

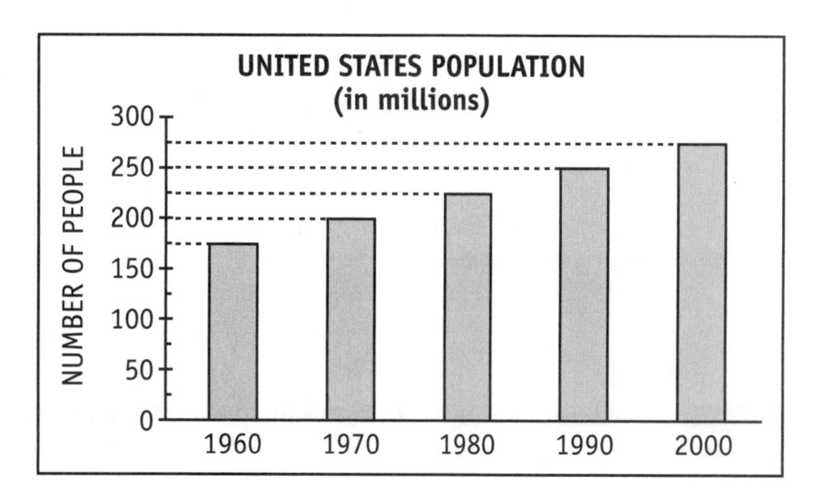

9. Complete the bar graph below to show the number of students enrolled in each grade at Chad Middle School.

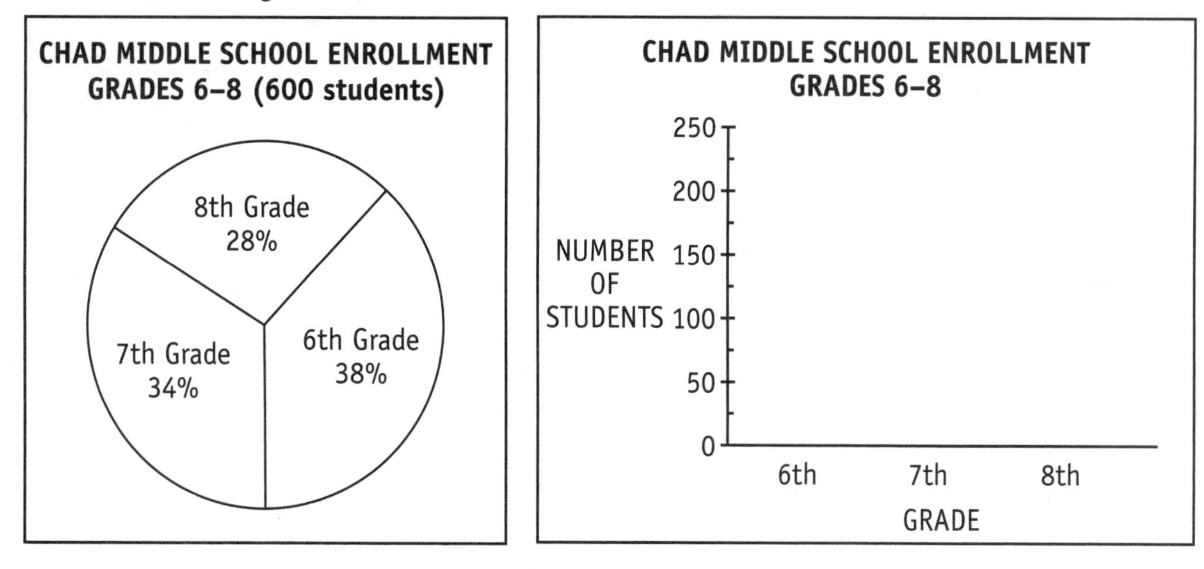

PROBABILITY

Introducing Probability

Probability is the study of **chance**—the likelihood of an event happening. Have you heard statements like these?

"There's a fifty percent chance of snow today."
"She has one chance in ten of getting that job!"
"Chances are five to one that the Cowboys will win."

The word *chance* indicates lack of control over what actually happens. Math reasoning skills help you answer many questions involving chance.

The pointer on the spinner can stop either on a shaded or an unshaded section. The pointer will not stop on a line.

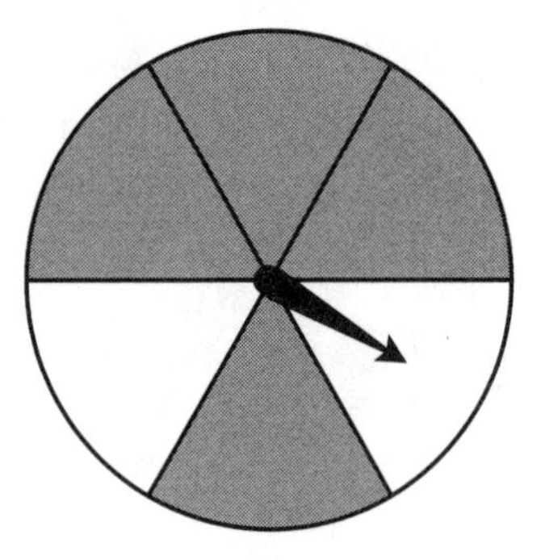

1. The spinner is divided into how many sections?

2. The pointer is most likely to stop on which type of section (shaded or unshaded)?

3. Suppose you spin the pointer 100 times. Circle the phrase that best describes the number of times the pointer is likely to stop on a shaded section.

   fewer than 50 *exactly 50* *more than 50*

Jerri remembers that 4, 6, and 9 are the three numbers of her locker combination. But she can't remember the exact combination.

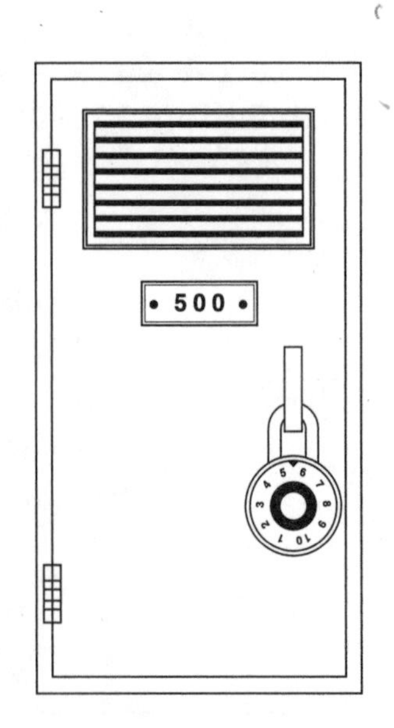

4. List the possible combinations for Jerri's locker.

5. Circle the phrase that describes Jerri's chance of choosing the correct combination on her first try.

   1 chance in 3 *1 chance in 6* *1 chance in 9*

6. Suppose it takes Jerri 30 seconds to try each combination. What is the greatest length of time it could take Jerri to open her locker?

The dart board has three scoring rings. Assume each tossed dart is equally likely to land *on any point on the board.*

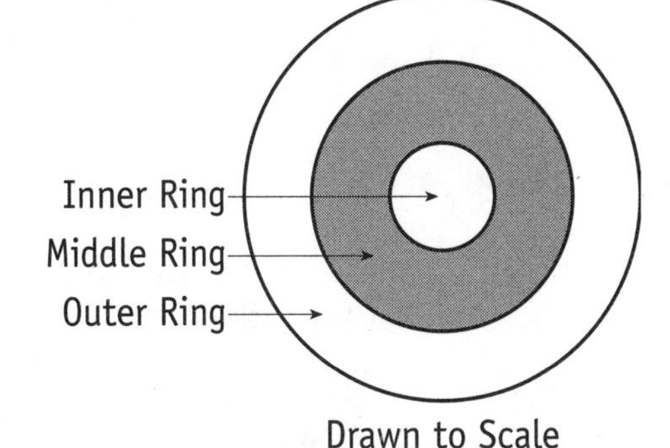

Drawn to Scale

7. a. On which ring is a dart most likely to land? Circle your answer.

inner ring *middle ring* *outer ring*

b. Briefly tell why you chose your answer in part a.

One of the three bags at the right contains candy. The other two contain sand.

8. Suppose you choose a bag without looking. Circle the phrase that best describes your chance of choosing the bag that contains the candy.

not very likely *very likely* *a sure thing*

9. Suppose on your first try you get a bag of sand. Circle the phrase that tells your chance of choosing the bag that contains the candy on your second try?

1 chance in 1 *1 chance in 2* *1 chance in 3*

10. Suppose on your second try you also get a bag of sand. How would you describe your chance of choosing the bag that contains the candy on your third try?

11. Colin puts cards with the numbers 1 to 40 written on them in a bag. Maggie draws a card without looking. Which is Maggie most likely to draw? A card that is

divisible by 3? *divisible by 4?* *divisible by 5?*

Give the reason for your answer.

Expressing Probability as a Number

Probability can be given as **chance,** such as 1 chance in 5, or as a number.

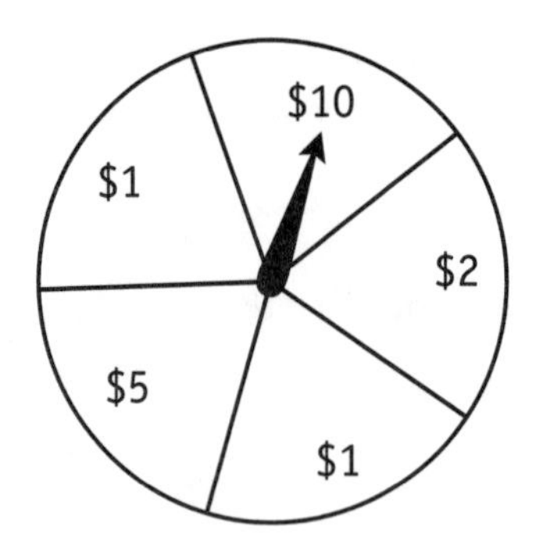

- a *fraction* between 0 and 1, such as $\frac{1}{5}$
- a *percent* between 0% and 100%, such as 20%

Probability tells only the *likelihood of an event happening.* It does not tell for sure what will happen.

The spinner is divided into five equal sections. On each spin, the pointer is equally likely to stop in any of the five sections. (Assume the pointer will not stop on a line.) You can ask, "What is the probability that the pointer will stop on $10?"

In the study of probability, each possible event (result) is called an **outcome.** A particular outcome, such as the pointer stopping on $10, is called a **favorable outcome.** The probability of a favorable outcome is written as a fraction.

$$\text{probability of a favorable outcome} = \frac{\text{number of favorable outcomes}}{\text{total number of possible outcomes}}$$

To write probability as a percent, change the fraction to a percent.

EXAMPLE 1 What is the probability that the pointer will stop on $10?

STEP 1 There are 5 possible outcomes. Of the 5, only 1 is the favorable outcome: $10. There is 1 chance out of 5 of the pointer stopping on $10.

STEP 2 Write the probability as a fraction.

$$\frac{\text{favorable outcomes}}{\text{possible outcomes}} = \frac{1}{5}$$

ANSWER: The probability that the pointer will stop on $10 is $\frac{1}{5}$, or **20%.**

On the average, only 1 spin out of 5 will stop on $10.

EXAMPLE 2 What is the probability that the pointer will stop on $1?

STEP 1 Of the 5 possible outcomes, 2 are favorable outcomes: $1 sections. There are 2 chances out of 5 of the pointer stopping on $1.

STEP 2 Write the probability as a fraction.

$$\frac{\text{favorable outcomes}}{\text{possible outcomes}} = \frac{2}{5}$$

ANSWER: The probability that the pointer will stop on $1 is $\frac{2}{5}$, or **40%.**

On the average, 2 spins out of 5 will stop on $1.

Write each probability as a fraction *and* as a percent. Reduce the fraction if possible.

1. a. 2 favorable outcomes out of 4 possible outcomes

_____ _____
fraction percent

b. 3 favorable outcomes out of 8 possible outcomes

_____ _____
fraction percent

c. 6 favorable outcomes out of 9 possible outcomes

_____ _____
fraction percent

For problem 2, refer to the spinner on page 90. Write each probability as a fraction.

2. What is the probability that the pointer will stop on

a. \$2?

b. \$1?

c. a value of \$2 *or more?*

For problems 3 and 4, refer to the spinner at the right.

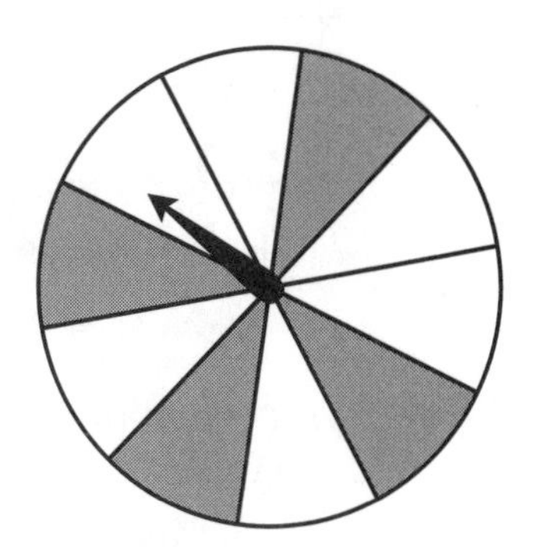

3. a. The spinner is divided into how many sections?

b. How many sections are shaded?

c. How many sections are unshaded?

4. a. What is the probability that the pointer will stop on a shaded section?

b. What is the probability that the pointer will stop on an unshaded section?

5. A number cube has six faces. Each face is equally likely to be up after the cube is tossed. With one toss of the cube, what is the probability of rolling

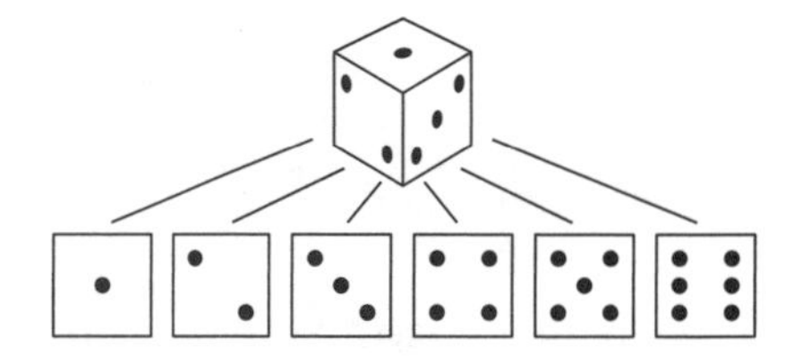

a. a 4?

b. an odd number (1, 3, or 5)?

c. an even number?

d. a number divisible by 3?

Probabilities from 0 to 1

Probabilities can have values from 0 to 1 (from 0% to 100%).

A Probability of 0

A probability of 0 (0%) means that an event cannot occur. The probability is 0 that the pointer on the spinner at the right will land on pink. There is no pink section!

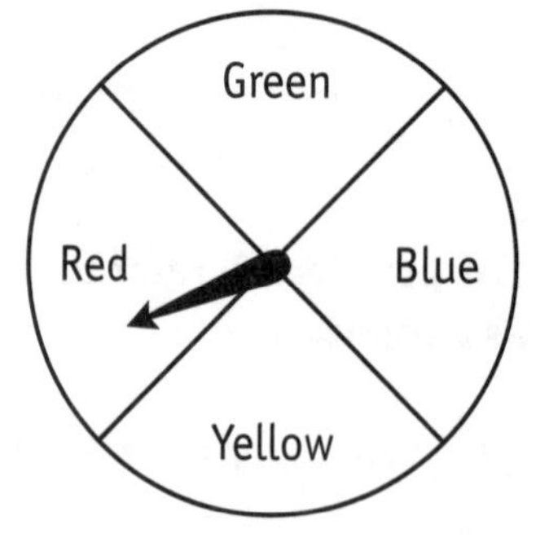

A Probability of 1

A probability of 1 (100%) means that an event will occur for sure. The probability is 1 that the pointer will land on one of the sections of the spinner. (Assume that the pointer does not land on a line and does not keep spinning forever. There is no other possibility!)

Probabilities between 0 and 1

Almost all probabilities are greater than 0 and less than 1.

- A probability less than $\frac{1}{2}$ (50%) means that an event is likely to happen less than half of the time—the smaller the probability, the less likely the event.
- A probability greater than $\frac{1}{2}$ (50%) means that an event is likely to happen more than half of the time—the larger the probability, the more likely the event.

For problems 1–8, write the probability.

1. On one spin, the pointer above will stop on gray. ______
2. On one spin, the pointer above will stop on either red *or* blue. ______
3. Somewhere in your state, a baby will be born today. ______
4. The next baby born in your town will be a girl. ______
5. The next baby born in your town will be a boy. ______
6. You will fly to the moon tomorrow. ______
7. A flipped penny will land heads up.
   (Assume that the penny doesn't land on its edge.) ______
8. Some place on Earth is experiencing night right now. ______

For problems 9 and 10, refer to the five cards shown below.

9. Suppose you randomly choose one card. What is the probability that the card you choose will be

   a. the king of hearts?

   b. a face card?

   c. a number card?

   d. a heart?

10. Suppose you randomly take one card, look at it, and return it. If you do this several times, which type of card (face card *or* number card) are you likely to get

    a. *less than* half of the time?

    b. *more than* half of the time?

For problems 11 and 12, refer to the coins which are placed in a bag.

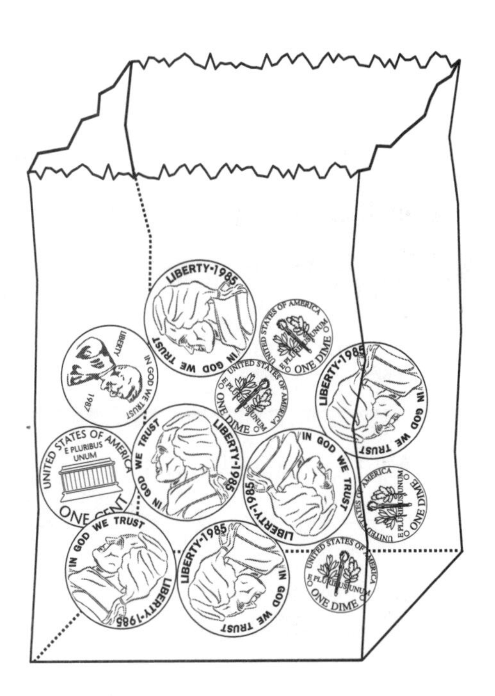

11. Suppose you randomly take one coin out of the bag. What is the probability that the coin will be a

    a. penny?

    b. nickel?

    c. dime?

    d. quarter?

12. Suppose you randomly take one coin out of the bag, look at it, and return it to the bag. If you do this several times, which coin(s) are you likely to get

    a. *less than* half of the time?

    b. *exactly* half of the time?

    c. *more than* half of the time?

Using Probability for Prediction

Probability can be used to make predictions. For example, if you roll a number cube 50 times, how many times are you likely to roll the number 6? You can't predict with certainty, but you can say how many 6s are *most likely.* To predict the number of 6s, multiply the number of rolls (50) by the probability of rolling a 6 on each roll.

EXAMPLE 1 If you roll a number cube 50 times, how many 6s will you most likely roll?

STEP 1 Write the probability of rolling a 6 on one roll.

$\frac{\text{favorable outcomes}}{\text{total outcomes}} = \frac{1}{6}$

STEP 2 Multiply 50 by $\frac{1}{6}$.

$50 \times \frac{1}{6} = \frac{50}{6} = \mathbf{8\frac{1}{3}}$

ANSWER: You will most likely roll **8** 6s.

EXAMPLE 2 If you flip a pair of coins 50 times, how many times will you most likely get two tails?

STEP 1 There are four possible results: H-H, H-T, T-H, and T-T. The probability of getting T-T (two tails) is 1 chance out of 4.

$\frac{\text{favorable outcomes}}{\text{total outcomes}} = \frac{1}{4}$

STEP 2 Multiply 50 by $\frac{1}{4}$.

$50 \times \frac{1}{4} = \frac{50}{4} = \mathbf{12\frac{1}{2}}$

ANSWER: You will most likely get **12** *or* **13** pairs of two tails.

For problem 1, refer to the number cube shown.

1. Suppose you roll a number cube 25 times.

   a. How many 6s will you most likely roll?

   b. How many even numbers will you most likely roll?

   c. How many numbers will you most likely roll that are divisible by 3?

For problem 2, refer to the coins at the right.

2. If you flip a pair of coins 50 times, how many times are you most likely to get 1 head and 1 tail? (See Example 2 above.)

For problem 3, refer to the spinner at the right.

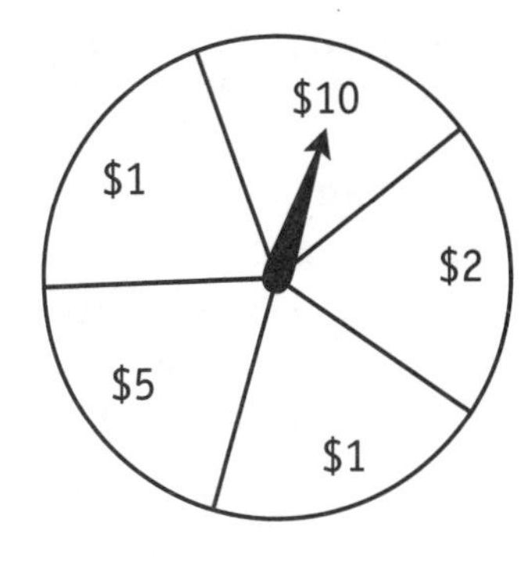

3. Suppose you spin the pointer 100 times.

   a. How many times is the pointer most likely to land on $10?

   b. How many times is the pointer most likely to land on $1?

For problem 4, refer to the dart board shown at the right. Assume that each ring of the board has the same area as each other ring.

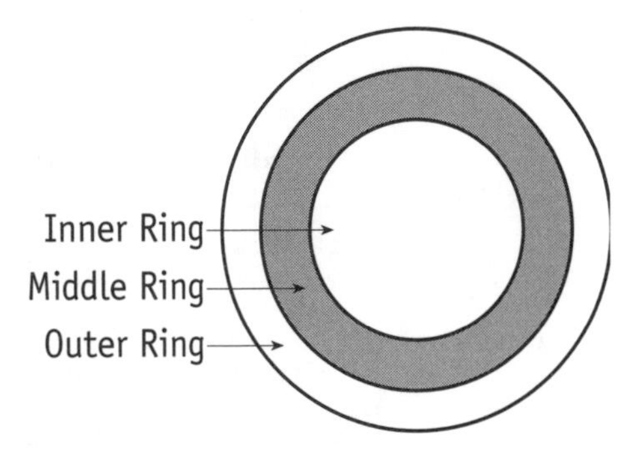

4. a. When Kelly tosses darts, each dart is equally likely to land in any of the three rings. If Kelly hits the board with 15 darts, how many are likely to hit the inner ring?

   b. If Kelly hits the board with only 10 darts, how many are likely to hit the outer ring?

5. Ellen, Denise, and Caren have a bag containing 12 marbles. Each takes a marble, records its color, and returns it to the bag. They each draw and return a marble to the bag a total of 12 times.

Ellen's Results	
Color	**Number**
red	1
blue	8
green	3

Denise's Results	
Color	**Number**
red	3
blue	5
green	4

Caren's Results	
Color	**Number**
red	2
blue	5
green	5

   a. What fraction of *all* 36 marbles resulted in marbles of each color?

   red: ______ blue: ______ green: ______

   b. The full bag contains 12 marbles. What is the *most likely* number of each color of marbles?

   red: ______ blue: ______ green: ______

Making a List to Predict Outcomes

For many probability problems, making a list of possible outcomes is the best first step. You then use your list to count total outcomes and favorable outcomes.

EXAMPLE Jenni, David, and Lauren have tickets to a play. They have three seats together in row G. Their ticket numbers are G6, G7, and G8. If each person randomly chooses a ticket, what is the probability that Lauren will sit next to David?

STEP 1 Make a list of all possible outcomes—all possible seating arrangements.

G6	G7	G8
Jenni	David	Lauren
Jenni	Lauren	David
David	Jenni	Lauren
David	Lauren	Jenni
Lauren	Jenni	David
Lauren	David	Jenni

STEP 2 Count the number of all possible seating arrangements. The total number of possible outcomes is 6.

STEP 3 Count the number of favorable outcomes—seating arrangements in which Lauren sits next to David. The number of favorable outcomes is 4. (A box is drawn around each of these.)

STEP 4 Write the probability as a fraction.

$$\frac{\text{favorable outcomes}}{\text{total outcomes}} = \frac{4}{6} = \frac{2}{3}$$

ANSWER: The probability that Lauren will sit next to David is $\frac{2}{3}$, or **$66\frac{2}{3}$%.**

Notice that the probability Lauren will *not* sit next to David is $\frac{2}{6} = \frac{1}{3}$, or **$33\frac{1}{3}$%.**

The sum of the two probabilities must equal 1 or 100%, since these are the only two possibilities.

Solve each problem. Write each probability either as a fraction or as a percent.

1. Look at the list of outcomes in the example on page 96.

   a. What is the probability that David will sit in seat G6?

   b. What is the probability that Jenni will sit to the right of, but not next to, Lauren?

2. A penny has a heads side and a tails side. Suppose you toss three pennies into the air and they land on the floor.

   a. Complete the table to show all possible combinations of tosses.

   b. What is the probability that all three pennies will land heads-up?

   c. What is the probability that two pennies will land heads-up and one will land tails-up?

Penny 1	Penny 2	Penny 3
heads	heads	heads
heads	heads	tails
heads	tails	
heads		
tails		
tails		
tails		
tails		

3. Amanda wins a dart game only if the sum of her two throws is 3 or less. Assume that each dart is equally likely to land in any section of the board.

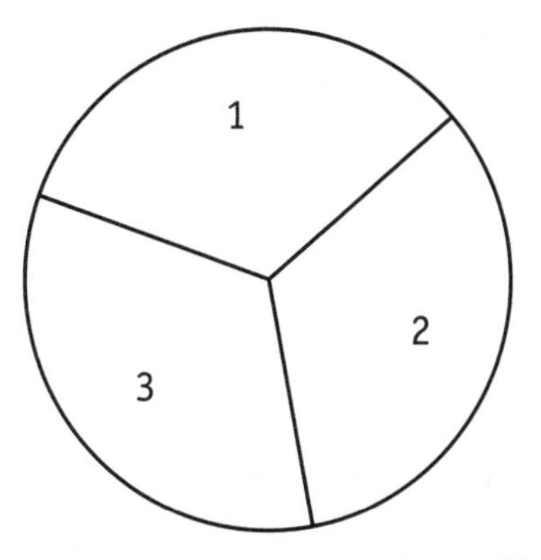

Dart 1	Dart 2	SUM Dart 1 + Dart 2
1	1	2
1	2	3
1	3	
2	1	
2		
2		
3		
3		
3		

   a. Complete the table to show all possible combinations of two darts.

   b. What sum is Amanda *most likely* to get?

   c. How many combinations win?

   d. What is the probability that Amanda will win?

Using a Tree Diagram

Another way to count outcomes is to draw a **tree diagram.** A tree diagram resembles a branching tree. Each path identifies a possible outcome.

EXAMPLE 1 Sandwich Heaven makes chicken, ham, and hamburger sandwiches. They also sell milk and soda. How many combinations of drinks and sandwiches are available?

You can draw a tree diagram to show each possible outcome.

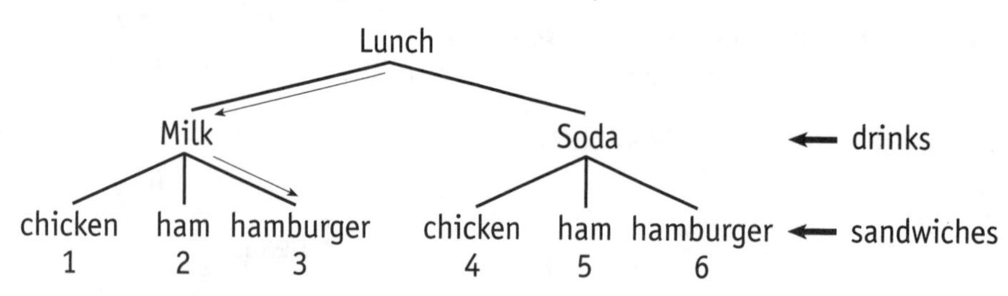

Each path (outcome) is numbered.

- Path 3 (small arrows) indicates a lunch of milk and a hamburger.
- Path 5 indicates a lunch of soda and a ham sandwich.

There are six separate paths—six outcomes or combinations.

ANSWER: 6 combinations

EXAMPLE 2 The Clothes Factory has XL sweatshirts on sale. The sweatshirts come only in blue or green. Each sweatshirt comes with either a zipper or a hood. And each may or may not have a store logo printed on it. How many combinations of sweatshirts are available?

Draw a tree diagram to show each possible outcome.

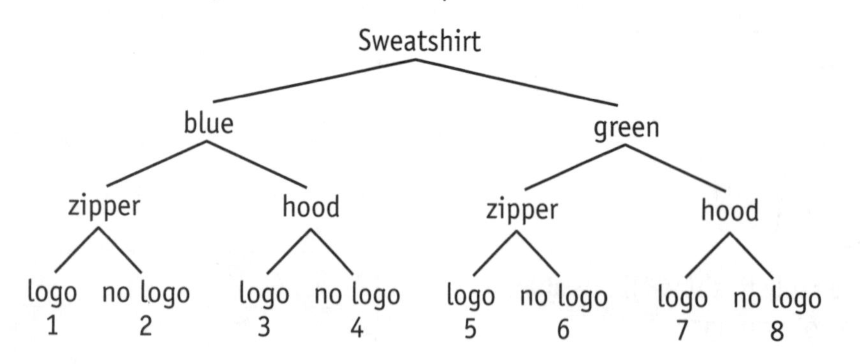

Again, each path is numbered. There are eight separate paths.

ANSWER: 8 combinations

For problems 1 and 2, refer to the examples on page 98.

1. At Sandwich Heaven, Greg lets Lola choose a drink and a sandwich for him. What is the probability that Greg will get a ham sandwich and milk?

2. The Clothes Factory has a gift box that contains one sweatshirt. The sales clerk tells you that it is equally likely to be any of the shirts on sale. What is the probability that the box contains a blue, hooded sweatshirt that does not contain a logo?

3. Jocelyn has three blouses: one plaid, one white, and one striped. She can mix and match the blouses with either her black or green skirt.

   a. Complete the tree diagram to show the number of outfits (blouse and skirt) Jocelyn can make.

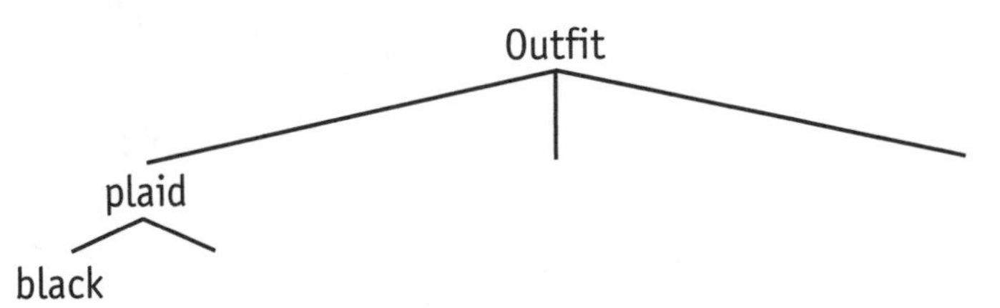

   b. How many outfits can Jocelyn make?

   c. If Jocelyn randomly chooses a blouse and skirt, what is the probability that she will choose the white blouse and the black skirt?

4. Marko has time to ride two more carnival rides. He must choose one ride from each of the two groups listed.

Group I	Group II
Octopus	Hurricane
Gravitron	Zipper
Python	Excaliber

   a. Complete the tree diagram to show the number of ride combinations Marko can choose.

   b. How many ride combinations are possible?

   c. Suppose Marko randomly chooses two rides. What is the probability he will choose the Gravitron and the Excaliber?

   d. Suppose Marko randomly chooses two rides but does not consider the Zipper which he doesn't like. What is the probability Marko will choose the Python and the Hurricane?

Basing Probability on Data

When you flip a coin, roll a number cube, or spin a pointer, you are basing probability on the laws of chance. You can also base probability on past performance or on data.

EXAMPLE 1 Claudia, a softball player, has gotten a hit 40 times out of 150 times at bat. What is the probability that Claudia will get a hit her first time at bat in tonight's game?

Claudia's past performance is used to predict her future performance. Up to now, Claudia has 40 *favorable outcomes* out of 150 *total outcomes.* Therefore, the probability that Claudia will get a hit her next time at bat is given by the following fraction.

$$\frac{\text{favorable outcomes}}{\text{total outcomes}} = \frac{40}{150} = \mathbf{\frac{4}{15}}$$

ANSWER: The probability is $\frac{4}{15}$ or about **27%** that Claudia will get a hit her first time at bat tonight.

EXAMPLE 2 The table below shows the types of cars sold at Motor City for the last 50 sales. What is the probability that the next car sold will be a Japanese car?

Type	Number
Domestic	24
European	6
Japanese	20

Of the last 50 cars sold, 20 were Japanese. Thus, 20 is the number of *favorable outcomes* and 50 is the number of *total outcomes.*

$$\frac{\text{favorable outcomes}}{\text{total outcomes}} = \frac{20}{50} = \mathbf{\frac{2}{5}}$$

ANSWER: The probability is $\frac{2}{5}$ or **40%** that the next car sold will be a Japanese car.

For problem 1, refer to Example 1 above.

1. a. Suppose Claudia bats only once in tonight's game. How would you describe her chances of getting a hit?

not likely *very likely* *a certainty*

b. Suppose Claudia bats 8 times in tonight's game. How many hits is she most likely to get?

For problem 2, refer to Example 2 above.

2. a. How many of the next 100 cars that Motor City sells are most likely to be European cars?

b. How many of these next 100 cars are most likely to be *either* domestic cars *or* Japanese cars?

Solve each problem. Write each probability as a fraction or as a percent.

3. Manuel, a basketball player, has made 30 free-throw shots out of his last 45 attempts.

a. What is the probability that Manuel will make his next free-throw?

b. How many free-throws is Manuel most likely to make out of his next 20 tries?

4. Sheila took two turns throwing darts. On each turn she threw five darts. Based on her results, what is the probability that the first dart Sheila throws on her third turn will score

a. 1 point?

b. 100 points?

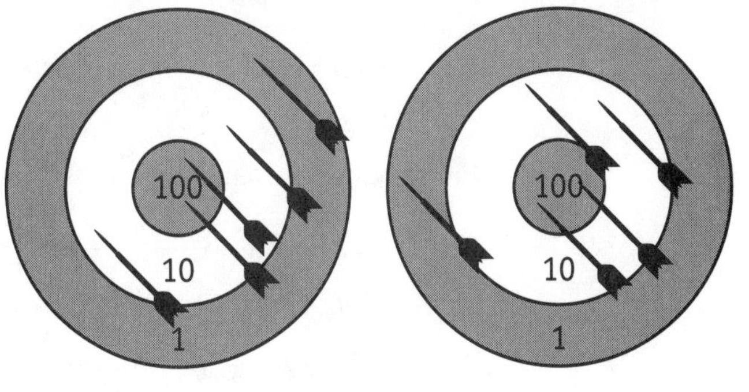

5. The graph below shows the ages of students at Granville Community College.

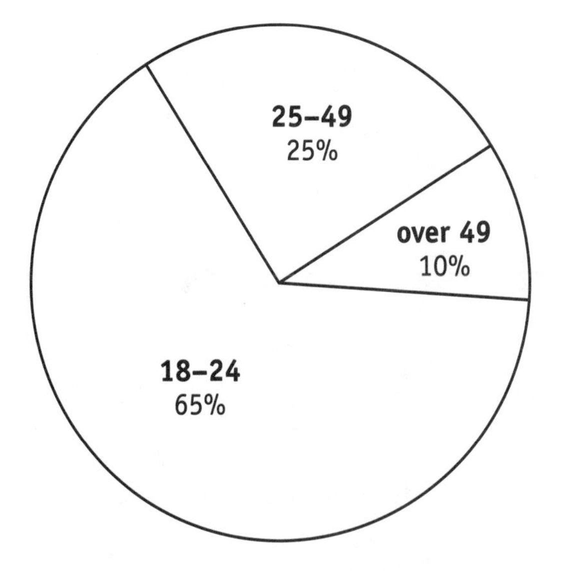

a. What is the probability that the next student who registers at Granville will be between the ages of 18 and 24?

b. Of the next 100 students who register at Granville, how many are most likely to be 50 or older?

6. On the first day of the children's pottery class, six girls and three boys registered.

a. What is the probability that the next child who registers will be a boy?

b. How many of the next 10 children who register will likely be girls?

7. Of the last 25 phone calls to the Reinke family, 15 have been for Tiffany.

a. What is the probability that the next phone call will *not* be for Tiffany?

b. How many of the next 15 phone calls will most likely *not* be for Tiffany?

Finding the Probability of Two Events

Sometimes you want to know the probability of two events happening in a row. This probability is found by multiplying the probability of the first event by the probability of the second.

Independent Events

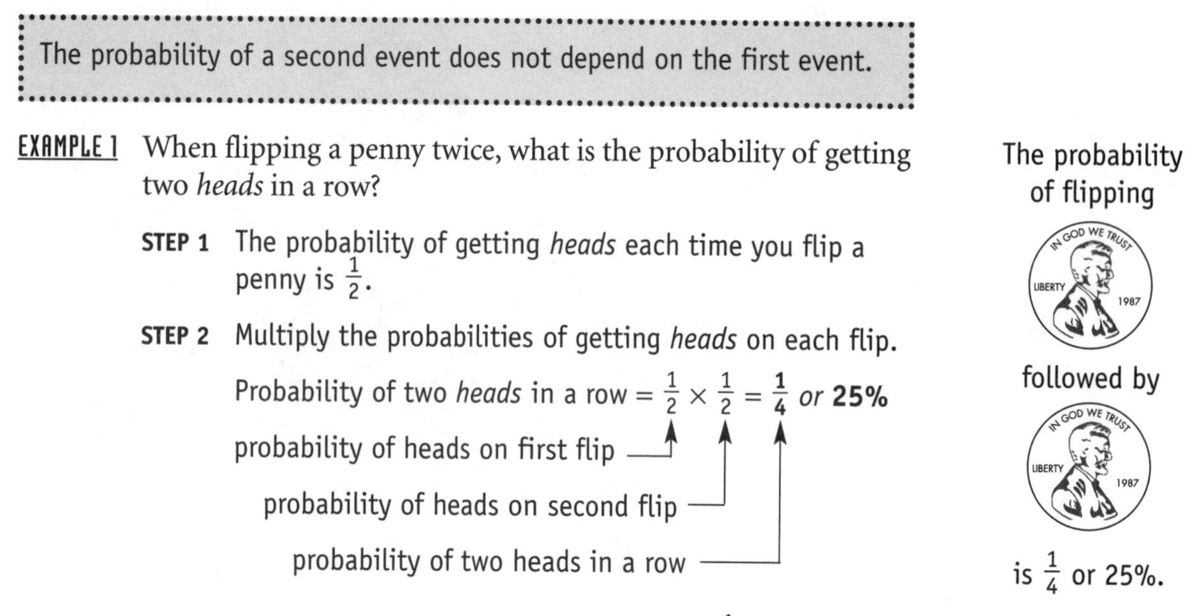

The probability of a second event does not depend on the first event.

EXAMPLE 1 When flipping a penny twice, what is the probability of getting two *heads* in a row?

STEP 1 The probability of getting *heads* each time you flip a penny is $\frac{1}{2}$.

STEP 2 Multiply the probabilities of getting *heads* on each flip.

Probability of two *heads* in a row = $\frac{1}{2} \times \frac{1}{2} = \frac{1}{4}$ *or* **25%**

probability of heads on first flip

probability of heads on second flip

probability of two heads in a row

The probability of flipping

followed by

is $\frac{1}{4}$ or 25%.

ANSWER: The probability of getting two heads in a row is $\frac{1}{4}$ *or* **25%.**

Dependent Events

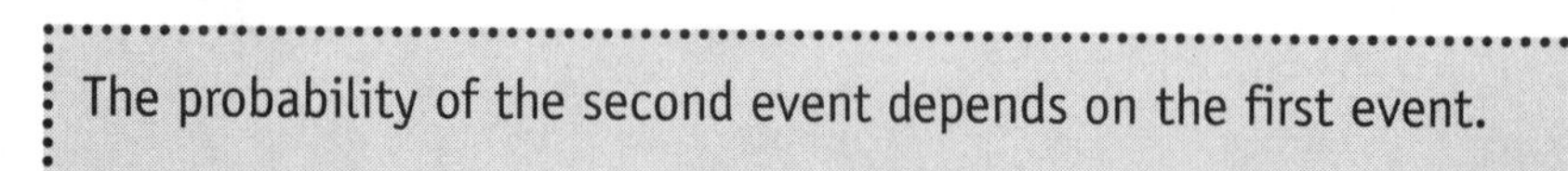

The probability of the second event depends on the first event.

EXAMPLE 2 If you randomly take two cards from the four cards shown, what is the probability that both cards will be face cards?

The probability of randomly drawing two face cards is $\frac{1}{6}$ or $16\frac{2}{3}$%.

STEP 1 When you take your first card, there are four possible outcomes, two of which are face cards. The probability of taking a face card as your first card is $\frac{2}{4}$ or $\frac{1}{2}$.

STEP 2 To calculate the second probability, *you assume that you do get a face card as your first card.* There are now three cards remaining, one face card and two number cards. The probability of taking a face card as your second card is $\frac{1}{3}$.

STEP 3 Multiply the first probability by the second. $\frac{1}{2} \times \frac{1}{3} = \mathbf{\frac{1}{6}}$

ANSWER: The probability of getting two face cards if $\mathbf{\frac{1}{6}}$ *or* **$16\frac{2}{3}$%.**

Express each probability as a fraction or as a decimal.

1. A number cube has six faces, numbered 1 to 6. Suppose you roll the cube twice.

   a. What is the probability of rolling a 6 on your first roll?

   b. What is the probability of rolling a 6 on your second roll?

   c. What is the probability of rolling two 6s in a row?

2. Suppose you randomly draw two cards from the cards shown below.

   a. What is the probability that the first card you draw will be a face card?

   b. What is the probability that both cards you draw will be face cards?

3. There are two questions on Colby's math test that he can't answer, so he decides to guess. Each question has five answer choices, one of which is correct.

   a. What is the probability that Colby can guess the answer to the first question?

   b. What is the probability that Colby can guess the answers to both questions?

4. The table below shows the transportation method used by students in Ms. Calder's class.

Transportation	Number
bus	18
car	6
walk	6

What is the probability that the next two students who enroll in Ms. Calder's class will come to school by bus?

5. The weather forecast says there is a 10% chance for snow tomorrow in Salem and a 25% chance for snow in Albany. What is the probability that it will snow in both Salem *and* Albany tomorrow?

6. Julian randomly chose a penny from the three shown. Then Matt chose one from the two that remained.

What is the probability that both boys chose heads?

Probability Review

Work each problem and check your answers. Correct any errors.

1. Lanni remembers that the last three digits of Shauna's phone number are 3, 6, and 9, but she can't remember the correct order of the digits!

   a. List all possible orders of the three digits.

   b. Circle the phrase that describes what chance Lanni has of guessing the correct order and being right on her first guess.

   1 chance in 2   1 chance in 4   1 chance in 6

For problems 2 and 3, refer to the spinner at the right.

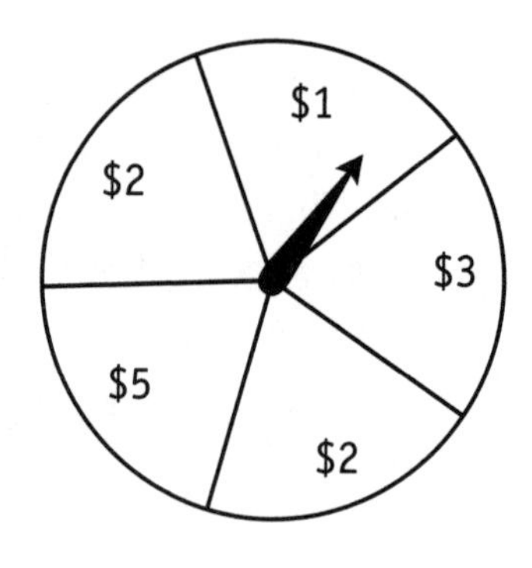

2. a. The spinner is divided into how many sections?

   b. On one spin, what is the probability that the pointer will land on the $5 section?

   c. On one spin, what is the probability that the pointer will *not* land on the $5 section?

3. a. On one spin, what is the probability that the pointer will land on a $2 section?

   b. On 50 spins, how many times is the pointer most likely to land on a $2 section?

   c. On 50 spins, how many times is the pointer most likely *not* to land on a $2 section?

4. A snack shop sells orange, cola, and lemon soft drinks. They also sell hot dogs, ham sandwiches, and chicken sandwiches.

   a. Complete the table to show all possible drink–sandwich combinations.

Drink	Sandwich
orange	hot dog

   b. If each combination is equally likely to be ordered, what is the probability that the next customer will order a ham sandwich and a cola?

5. At her school's family fun night, Marcella took two turns at the ball toss. On each turn, she tossed five balls. Based on her results, what is the probability that Marcella's first toss on her third turn will score

   a. 0 points?

   b. 5 points?

   c. 10 points?

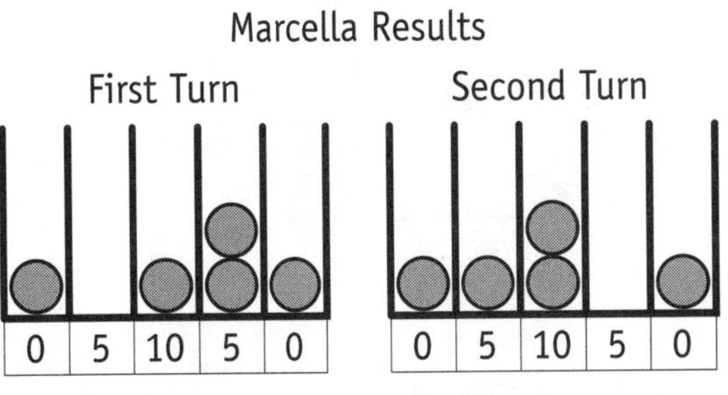

Hint: How many of each value (0, 5, and 10) did Marcella get in total?

6. Amanda's Restaurant sells four types of omelets for breakfast. The table shows the type of omelets ordered by the last 20 customers.

Type of Omelet	Number
Cheese	5
Ham	7
Vegetarian	2
Sausage	6

   a. What is the probability that the next omelet ordered will be cheese?

   b. What is the probability that the next two omelets ordered will *both* be cheese?

   c. What is the probability that the next two omelets ordered will *not* both be cheese? (**Hint:** Use your answer from part b.)

GEOMETRY AND MEASUREMENT

Lines

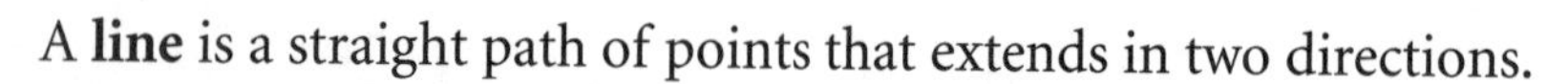

A **line** is a straight path of points that extends in two directions.

line

- **Parallel lines** run side by side and never cross.

parallel lines

- **Perpendicular lines** cross at a right angle (a corner angle).

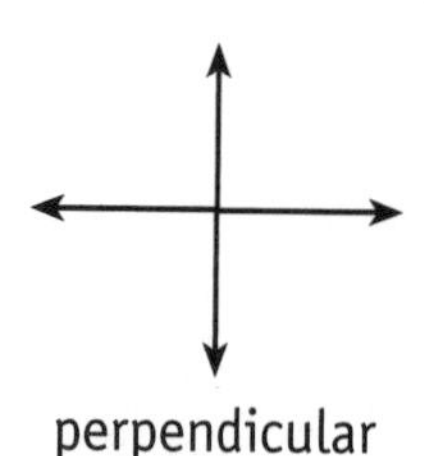

perpendicular lines

Did You Know . . . ?
A right angle is one-fourth of a circle and has a measure of 90°.

- A **horizontal line** runs left to right.

horizontal line

- A **vertical line** runs straight up and down.

vertical line

A **line segment** is a straight path of points that has two endpoints.

- A line segment is used to form the sides of a geometric figure such as a **triangle.**

line segment

triangle

A **ray** is a straight path of points that has one endpoint.

- An **angle** is formed by two rays that are joined at their endpoints.

ray

Line segments and rays are usually just referred to as *lines.*

angle

For problems 1–5, refer to the map at the right.

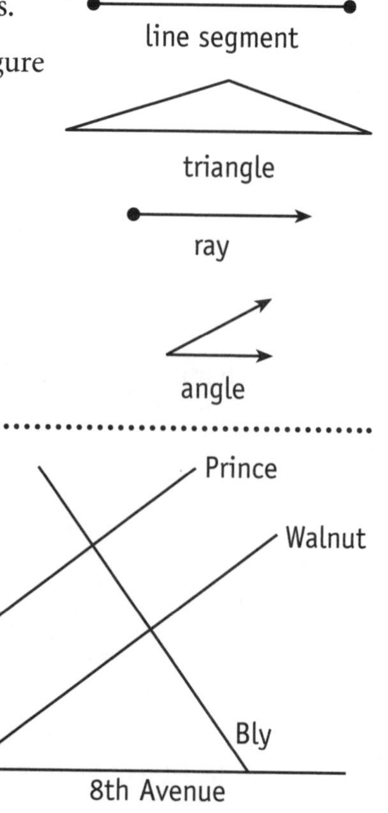

Which street

1. is parallel to Walnut?
2. is perpendicular to Prince?
3. runs vertically?
4. runs horizontally?
5. forms a 90° angle where it joins King Street?

Angles

Two rays that form an angle are called the **sides** of the angle. The point where the rays meet is called the **vertex.**

sides

vertex

Naming an Angle

An angle is named by a single letter or by three letters. In a three-letter name, the vertex letter is always the middle letter. (Numbers can also be used to name angles.)

∠C ∠x ∠RST

Measuring an Angle

The measure (size) of an angle is the opening between its sides. This opening is given in units called **degrees.** The symbol for degrees is °. A 45-degree angle is written as 45°.

Degrees can be thought of as part of a divided circle. A whole circle contains 360°.

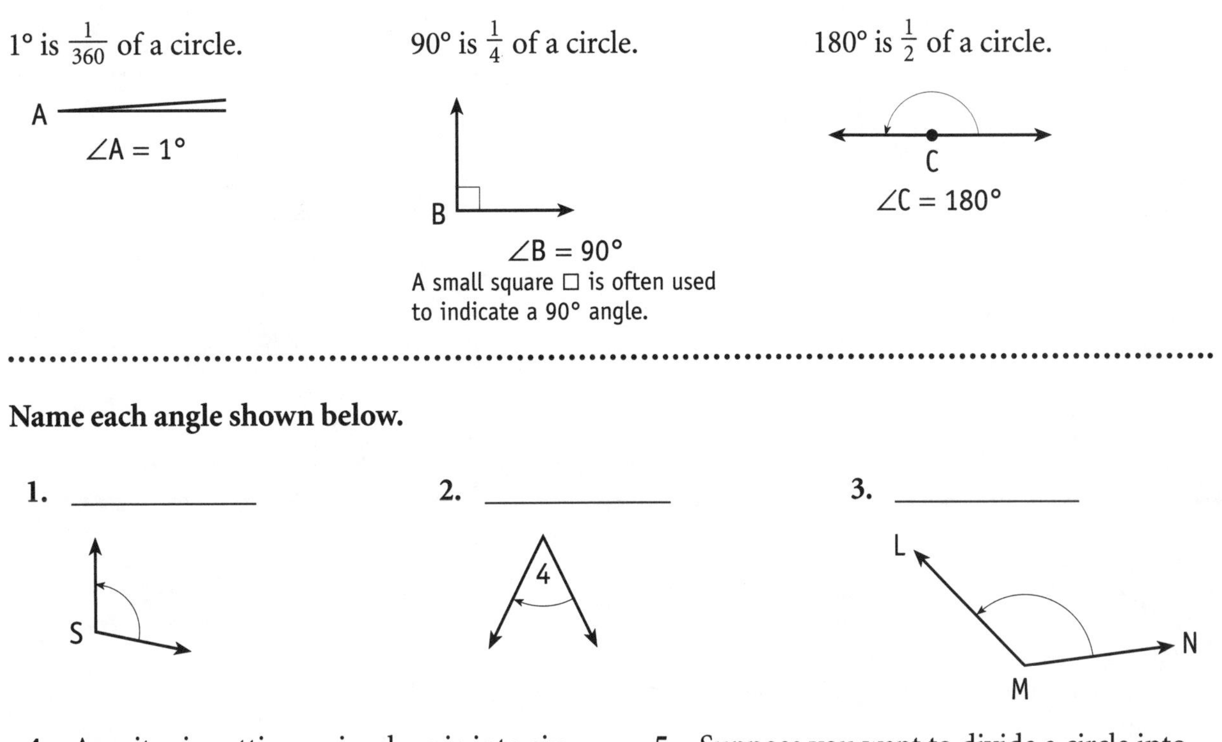

4. A waiter is cutting a circular pie into six equal slices.

a. At what angle should he cut the sides of each piece? (**Hint:** 360° ÷ ?)

b. Draw this angle.

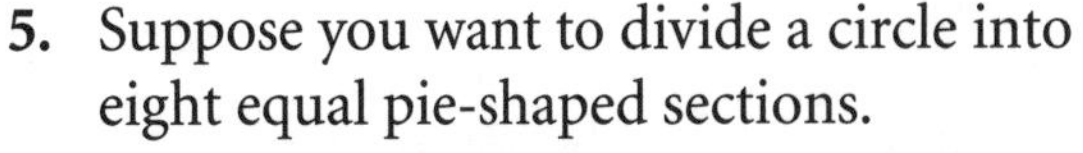

5. Suppose you want to divide a circle into eight equal pie-shaped sections.

a. At what angle will the sides of each section meet? (**Hint:** 360° ÷ ?)

b. Draw this angle.

Types of Angles

An angle can be classified by its size.

Acute Angle

41°

greater than 0°
but less than 90°

Right Angle

90°

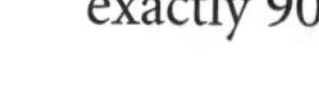

exactly 90°

Obtuse Angle

130°

greater than 90°
but less than 180°

Straight Angle

180°

exactly 180°

Reflex Angle

195°

greater than 180°
but less than 360°

Classify each angle as acute, right, obtuse, straight, or reflex.

1. ____________

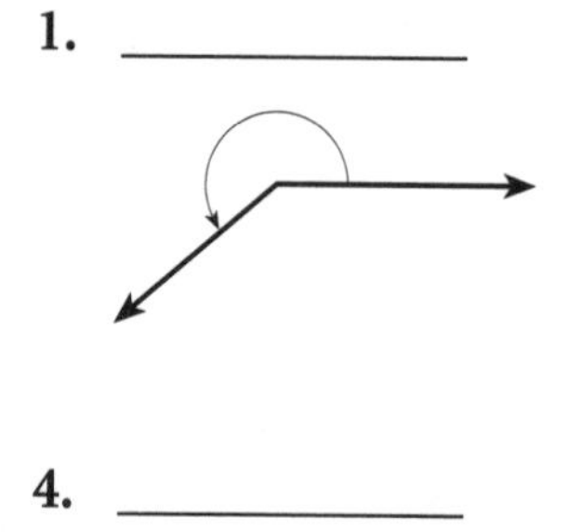

2. ____________

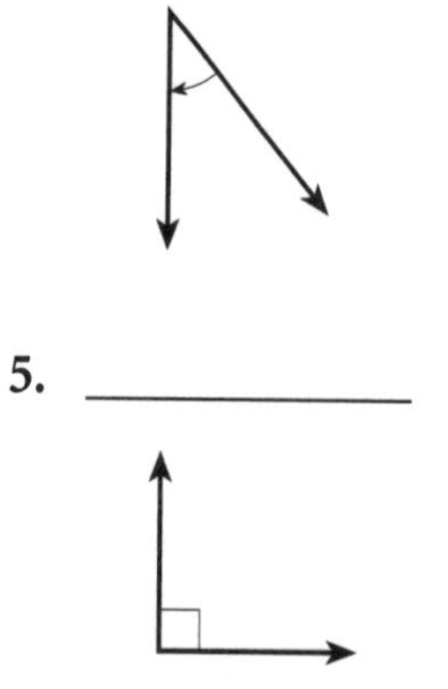

3. ____________

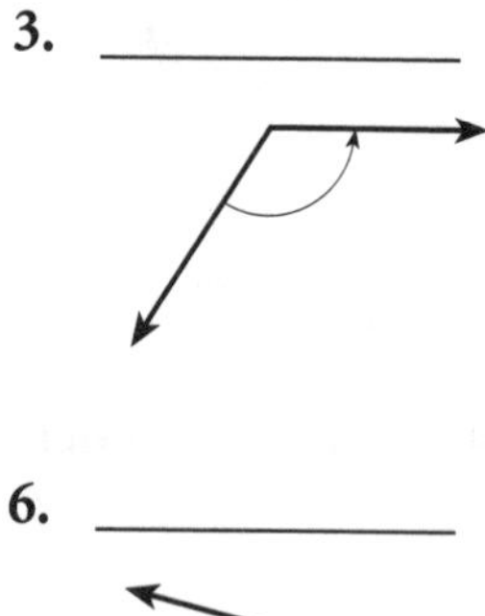

4. ____________

5. ____________

6. ____________

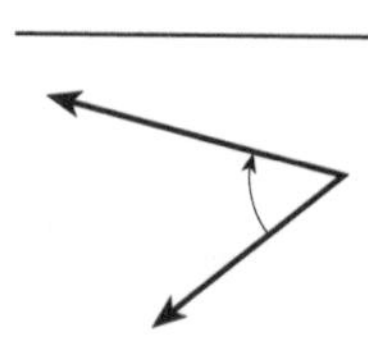

Draw an example of each type of angle. Estimate the measure of each angle.

7. acute
8. right
9. obtuse
10. straight
11. reflex

Pairs of Angles

Angles can be added to form a larger angle.

- Angles that add to 90° are called **complementary angles.**
- Angles that add to 180° are called **supplementary angles.**

Complementary Angles

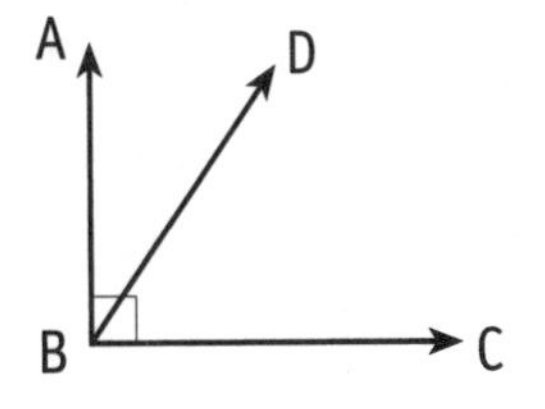

$\angle ABD + \angle DBC = 90°$

∠ABD and ∠DBC are complementary.
∠ABD is the complement of ∠DBC.

Supplementary Angles

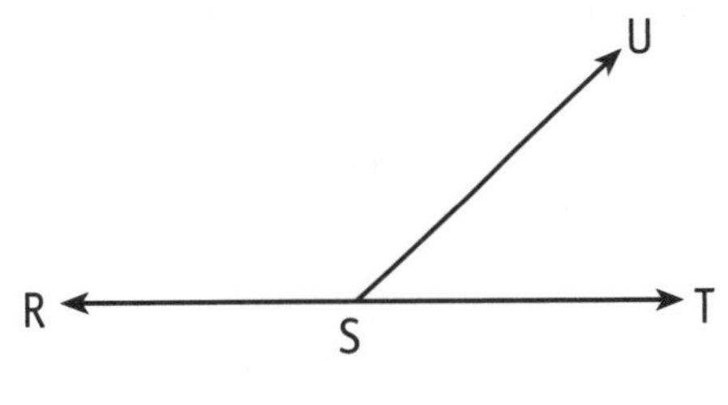

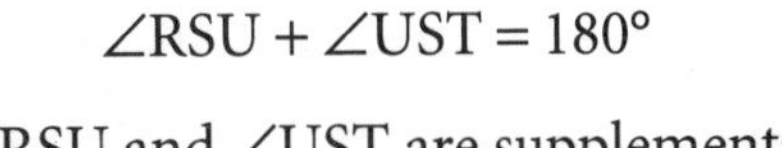

$\angle RSU + \angle UST = 180°$

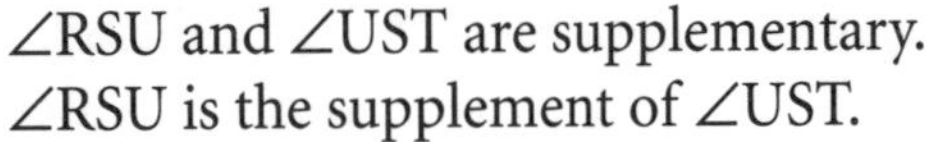

∠RSU and ∠UST are supplementary.
∠RSU is the supplement of ∠UST.

EXAMPLE Angle ABC is a right angle. What is the measure of ∠DBC if ∠ABD = 32°?

$\angle DBC = 90° - 32° = \mathbf{58°}$

ANSWER: ∠DBC = 58°

A, D, 32°, ?, B, C

$\angle ABD + \angle DBC = 90°$

Find the measure of each angle.

1. ∠ABD = _______ **2.** ∠QST = _______ **3.** ∠CBD = _______

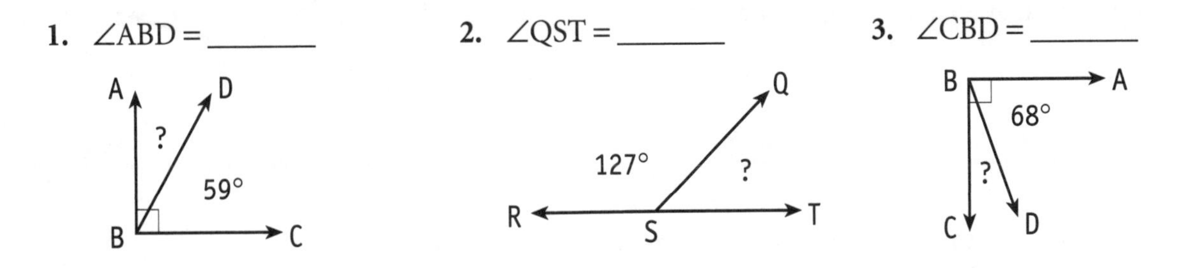

4. Draw an angle (∠LMO) that is complementary to ∠LMN. *Estimate* the measure of ∠LMO.

∠LMO ≈ _______

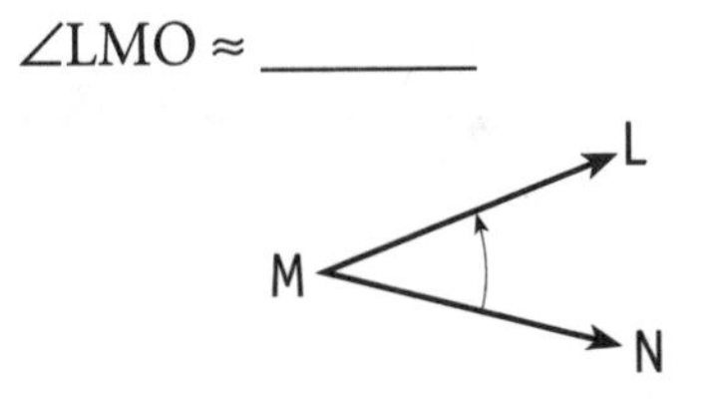

5. Draw an angle (∠ABD) that is supplementary to ∠ABC. *Estimate* the measure of ∠ABD.

∠ABD ≈ _______

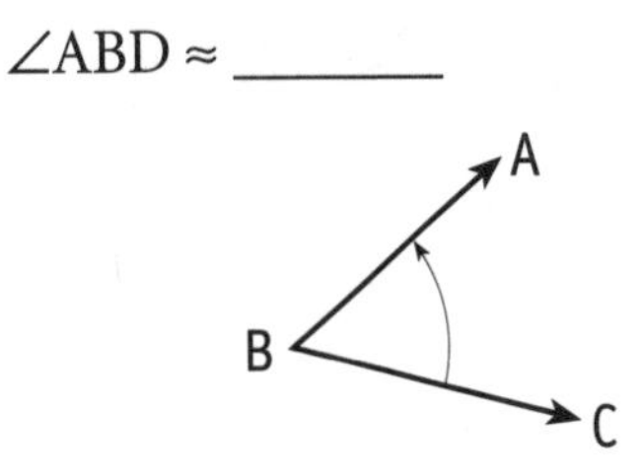

Angles Formed by Intersecting Lines

Four angles are formed when two lines cross (intersect).

Adjacent Angles

Angles that share a side are called **adjacent angles.**

In the drawing at the right, there are four pair of adjacent angles: $\angle a$ and $\angle b$; $\angle b$ and $\angle c$; $\angle a$ and $\angle d$; and $\angle c$ and $\angle d$.

- Adjacent angles formed by intersecting lines are supplementary.

$\angle a + \angle b = 180°$ $\angle b + \angle c = 180°$
$\angle a + \angle d = 180°$ $\angle c + \angle d = 180°$

Vertical Angles

Angles that are opposite each other are called **vertical angles.**

- Vertical angles are equal.

$\angle a = \angle c$ and $\angle b = \angle d$

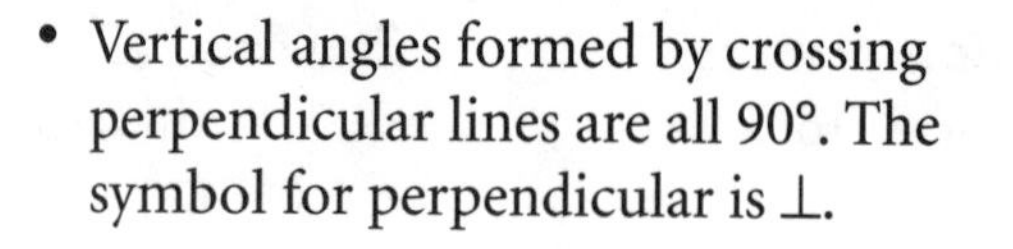

- Vertical angles formed by crossing perpendicular lines are all 90°. The symbol for perpendicular is ⊥.

A ⊥ B

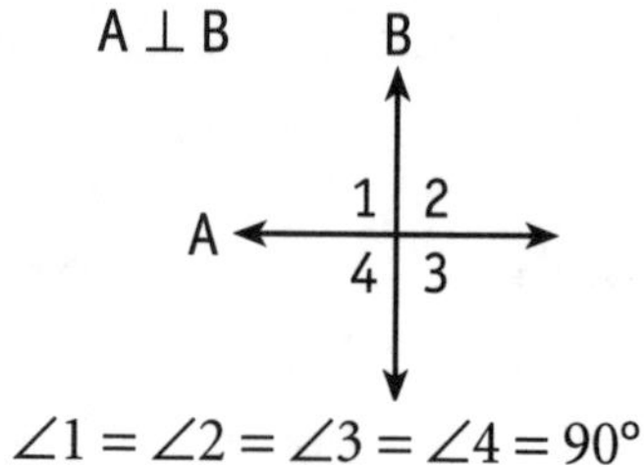

$\angle 1 = \angle 2 = \angle 3 = \angle 4 = 90°$

For problem 1, refer to the figure at the right.

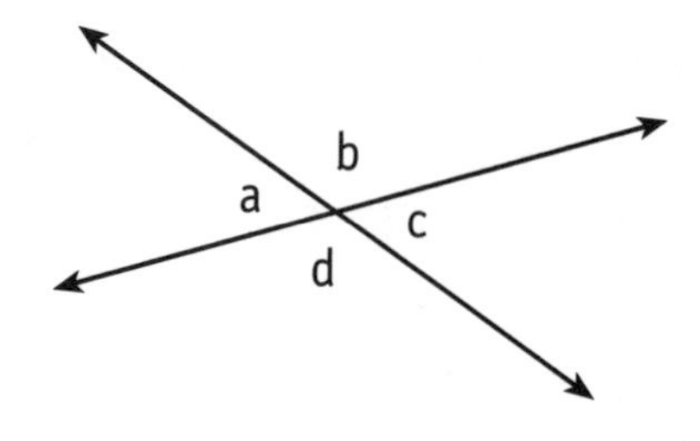

1. **a.** Which angle is vertical to $\angle b$?

   b. Name an angle that is supplementary to $\angle c$.

   c. If $\angle a = 55°$, what is the measure of each other angle?

   $\angle b =$ ______ $\angle c =$ ______ $\angle d =$ ______

For problem 2, refer to the street map. In this map, King Blvd. is perpendicular to 12th Ave.

2. **a.** Which two angles have measures of 90°?

   b. Which angle has the same measure as $\angle c$?

   c. Which angle is vertical to $\angle f$?

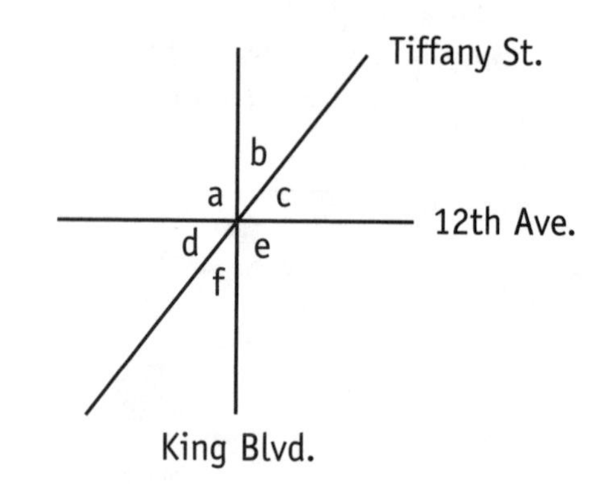

Parallel Lines Cut by a Transversal

A **transversal** is a third line that crosses two parallel lines. The symbol for parallel is ||. At the right, line C is parallel to line D, or C || D.

When two parallel lines are cut by a transversal,

- the four acute angles are equal
- the four obtuse angles are equal
- each acute angle is supplementary to each obtuse angle

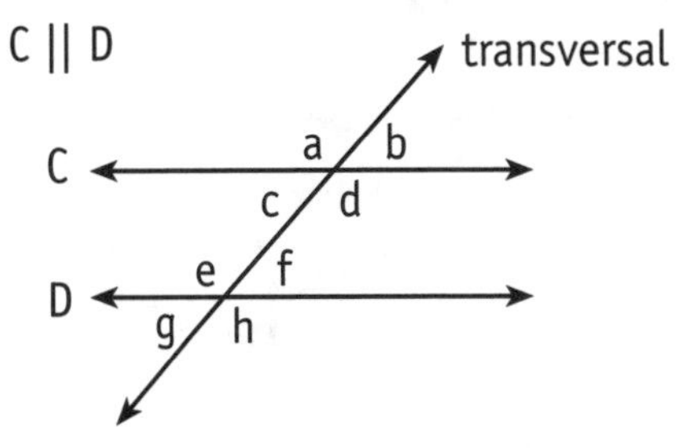

$\angle c = \angle b$ $\angle a + \angle b = 180°$
$\angle g = \angle f$ $\angle e + \angle f = 180°$
and so on and so on

For problems 3 and 4, find the measure of each angle as indicated.

3. A || B

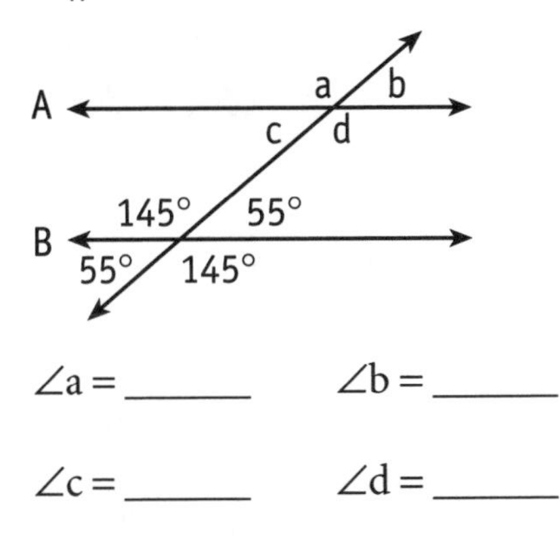

$\angle a =$ ______ $\angle b =$ ______

$\angle c =$ ______ $\angle d =$ ______

4. C || D

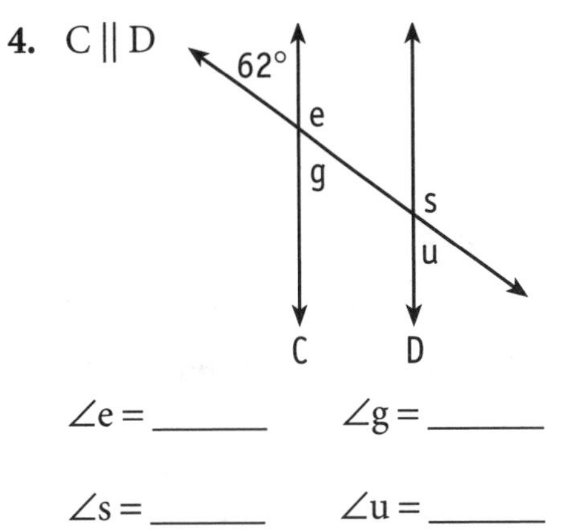

$\angle e =$ ______ $\angle g =$ ______

$\angle s =$ ______ $\angle u =$ ______

5. Use the information given on the street map at the right to determine the measures of $\angle x$ and $\angle y$.

$\angle x =$ ______

$\angle y =$ ______

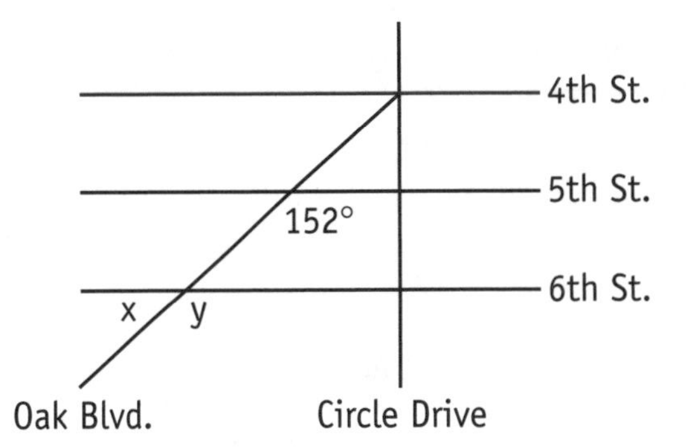

6. A parallelogram has two pairs of parallel lines. Use your knowledge of angles to find the sum of the four angles of a parallelogram.

$\angle A + \angle B + \angle C + \angle D =$ ______

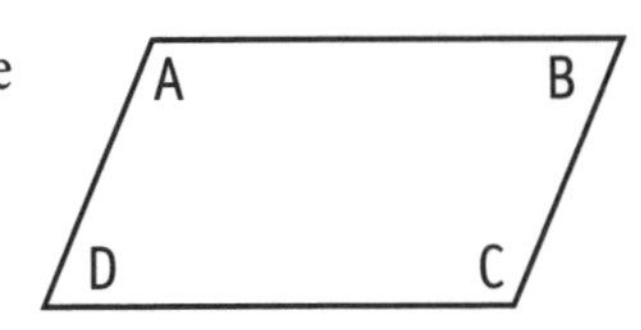

Properties of Triangles

Triangles can be drawn in a variety of shapes, but all share the properties listed below.

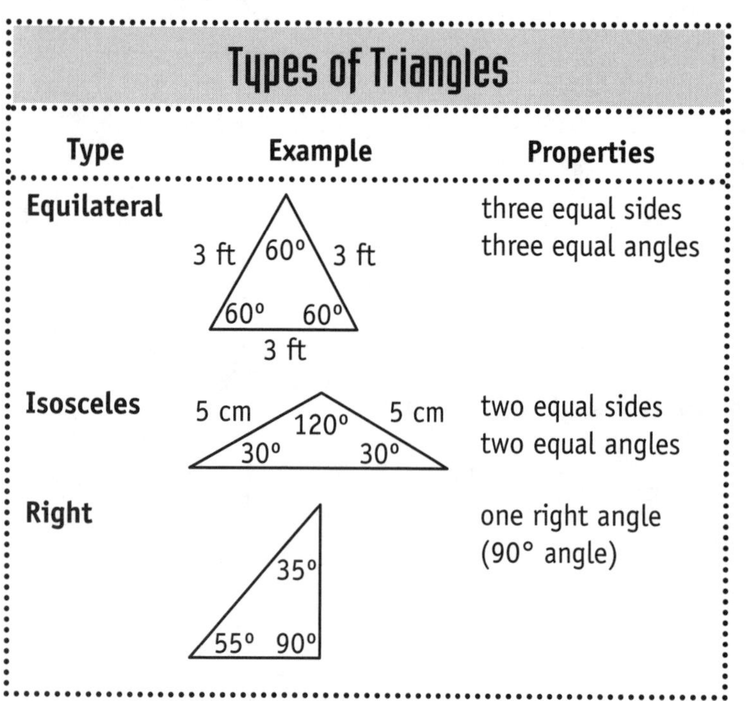

Types of Triangles

Type	Example	Properties
Equilateral	3 ft, 3 ft, 3 ft; 60°, 60°, 60°	three equal sides three equal angles
Isosceles	5 cm, 5 cm; 120°, 30°, 30°	two equal sides two equal angles
Right	35°, 55°, 90°	one right angle (90° angle)

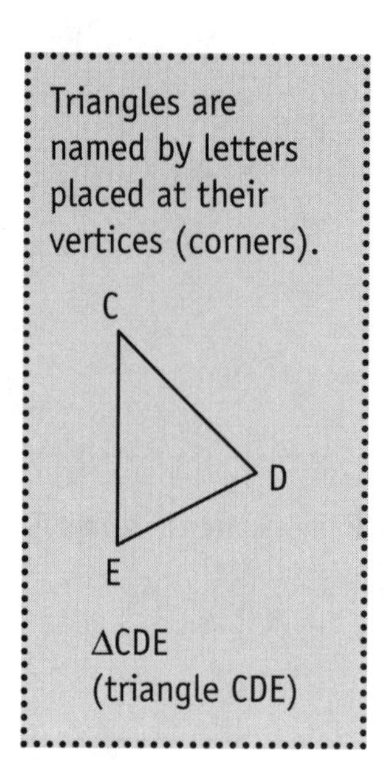

Triangles are named by letters placed at their vertices (corners).

ΔCDE
(triangle CDE)

Important Properties of Triangles

- The sum of the three angles of a triangle is 180°.
- In an isosceles or equilateral triangle, equal sides are opposite equal angles.
- In an isosceles or equilateral triangle, equal angles are opposite equal sides.
- The longest side of any triangle is opposite the largest angle.

EXAMPLE 1 In ΔABC, what is the measure of ∠C?

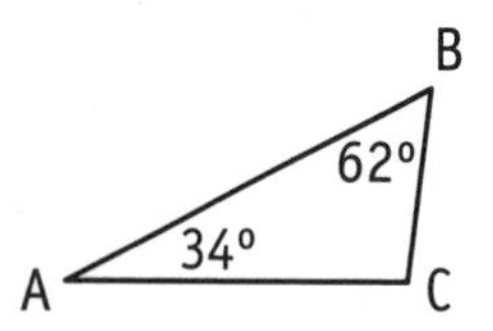

STEP 1 Add ∠A + ∠B. 34° + 62° = 96°

STEP 2 Subtract 96° from 180°.

180° − 96° = **84°**

ANSWER: ∠C = **84°**

EXAMPLE 2 Triangle XYZ is an isosceles triangle. If ∠Y measures 80°, what is the measure of ∠X?

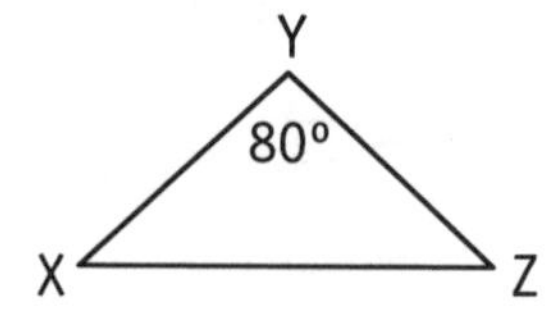

STEP 1 Subtract. 180° − 80° = 100°

(∠X + ∠Z = 100°)

STEP 2 Divide 100° by 2.
100° ÷ 2 = **50°**

(Since ∠X = ∠Z; ∠X = 100° ÷ 2)

ANSWER: ∠X = **50°**

Classify each triangle as equilateral, isosceles, or right.

1. ____________ **2.** ____________ **3.** ____________ **4.** ____________

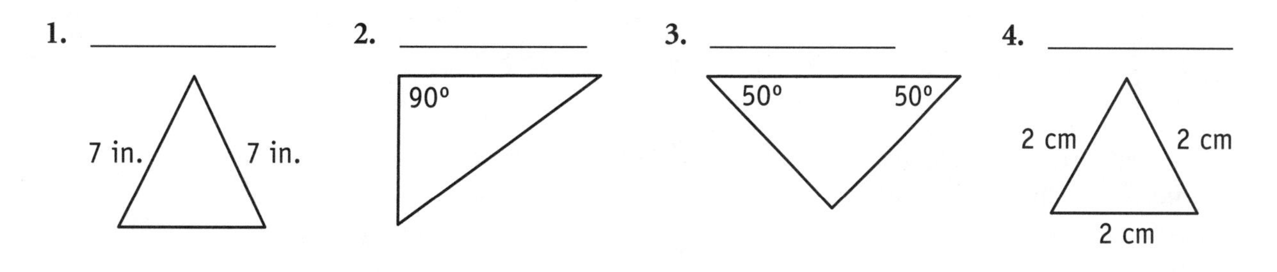

Find the measure of each unlabeled angle.

5. ∠B = ______ **6.** ∠D = ______ **7.** ∠Y = ______ **8.** ∠T = ______

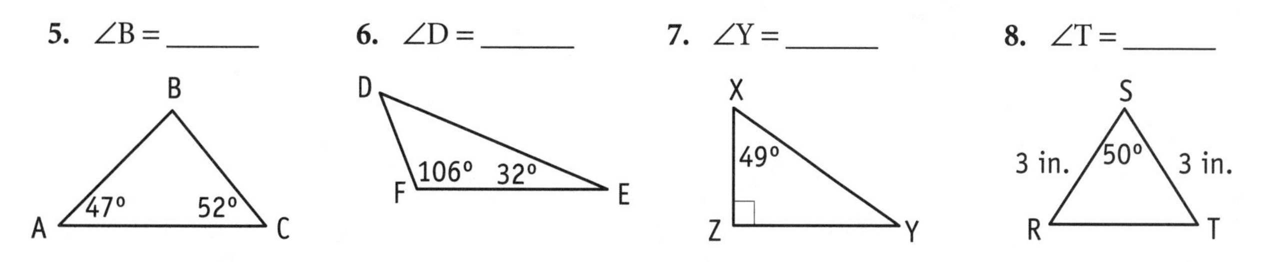

Answer each question.

9. Can a right triangle have two right angles? If so, draw an example.

Yes No

10. Can a triangle have three acute angles? If so, draw an example.

Yes No

11. Can a triangle have two obtuse angles? If so, draw an example.

Yes No

12. Can a triangle have three 60° angles? If so, draw an example.

Yes No

13. What type of triangle can be formed from five toothpicks if the toothpicks are placed end to end? Draw the triangle.

14. What type of triangle can be formed from six toothpicks if the toothpicks are placed end to end? Draw the triangle.

Perimeter

Perimeter is the distance around a plane figure. The symbol for perimeter is *P*.

A **polygon** is a plane figure made up of three or more line segments. To find the perimeter of a polygon, add the lengths of its sides.

Formulas for the perimeters of four common polygons are shown below.

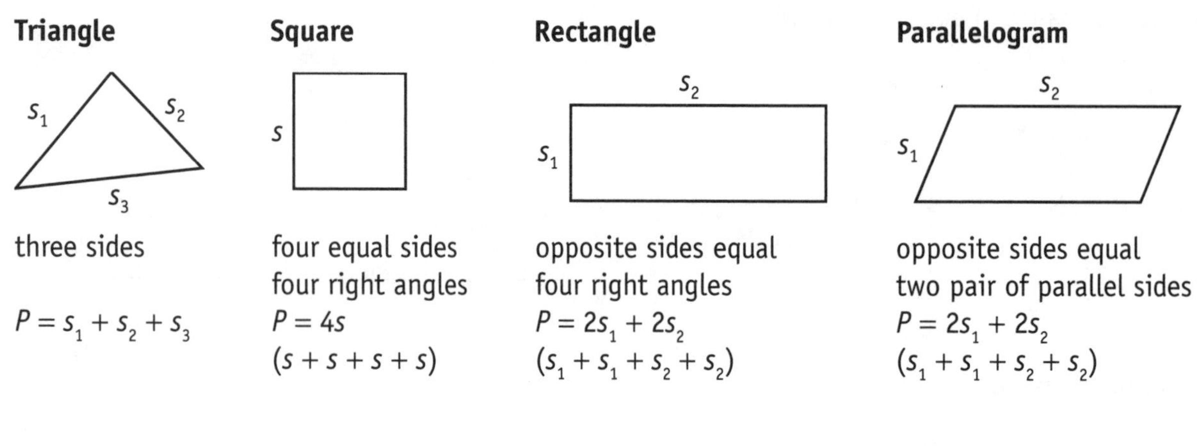

EXAMPLE What is the perimeter of a rectangle that is 4 yards wide and 9 yards long?

To find the perimeter, add the lengths of the four sides *or* use the perimeter formula for a rectangle.

$P = 2s_1 + 2s_2 = (2 \times 4) + (2 \times 9)$ (equal to $4 + 4 + 9 + 9$)
$= 8 + 18$
$=$ **26 yards**

ANSWER: 26 yards

> **Did You Know . . . ?**
> A *regular polygon* is a polygon in which all angles and all sides are equal.
>
> Which of the four polygons shown above is a regular polygon?

Find the perimeter of each figure.

1. $P =$ ____________

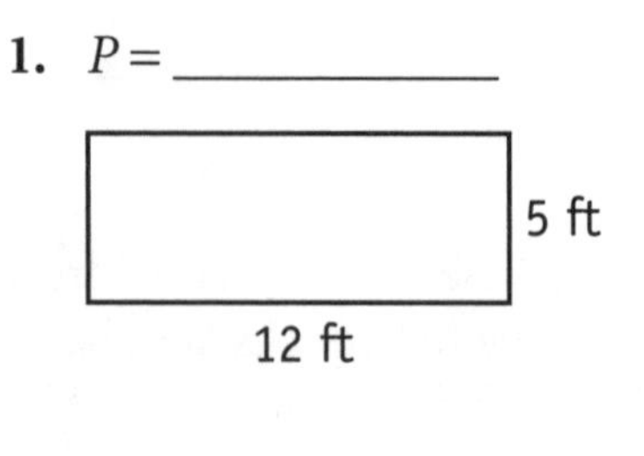

2. $P =$ ____________

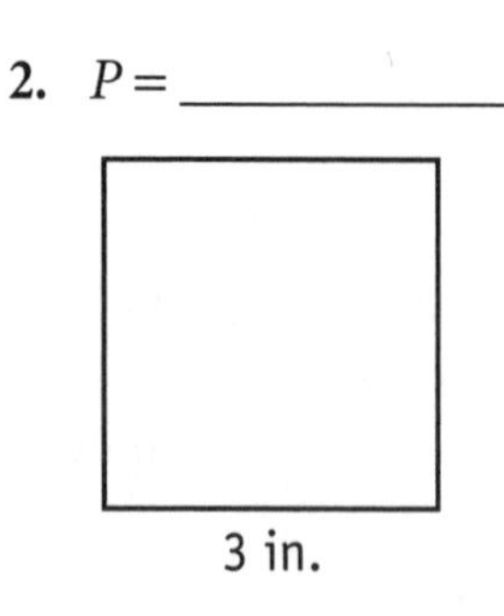

3. $P =$ ____________

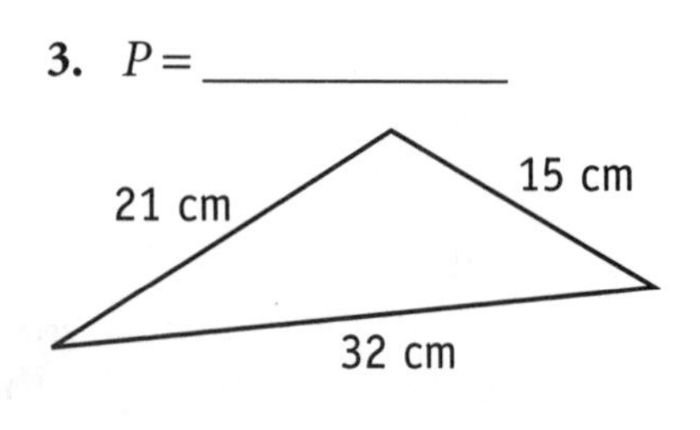

Circumference of a Circle

A **circle** is a figure in which all points are at an equal distance from the center. The perimeter of a circle is called the **circumference** (C).

- The **radius** (r) is the distance from the center to any point.
- The **diameter** (d) is the distance across a circle. The diameter is equal to twice the radius. $d = 2r$
- The **circumference** is given by the formula $C = \pi d$, where π (pronounced "pie") has the approximate value 3.14 or $\frac{22}{7}$.

Circle

Center

Diameter

Radius

$d = 2r$ or $r = \frac{1}{2}d$

To find the circumference of a circle, multiply the diameter by π.

EXAMPLE 1 What is the circumference (C) of a circle that has a diameter of 14 inches?

To find C, multiply 14 by $\frac{22}{7}$.

$C = \pi d \approx 14 \times \frac{22}{7} = \frac{\overset{2}{\cancel{14}}}{1} \times \frac{22}{\cancel{7}_1} = \mathbf{44}$

ANSWER: about 44 inches

(**Note:** Use $\pi \approx \frac{22}{7}$ for a radius or diameter that is divisible by 7.)

EXAMPLE 2 What is the circumference (C) of a circle that has a diameter of 2.4 feet?

To find C, multiply 2.4 by 3.14.

$C = \pi d \approx 2.4 \times 3.14 = \mathbf{7.536} \approx \mathbf{7.5}$

ANSWER: about 7.5 feet

(**Note:** Use $\pi \approx 3.14$ when the radius or diameter is a decimal.)

Find the circumference of each circle. Round each decimal answer to the nearest tenth.

1. $C =$ ____________ **2.** $C =$ ____________ **3.** $C =$ ____________

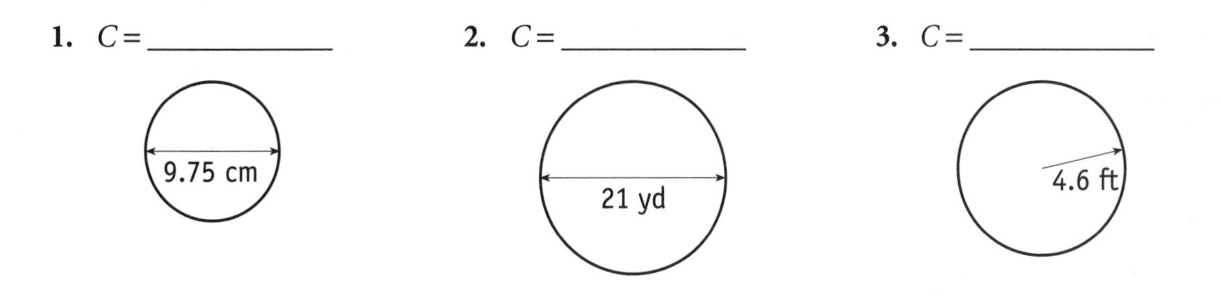

4. The diameter of a circular wading pool is 28 feet. What is the approximate distance around the outside edge of the pool?

5. The radius of the earth is about 4,000 miles. What is the approximate distance around the earth at the equator?

Area of Squares and Rectangles

Sample Area Unit
1 square foot

1 ft
1 ft

Area (A) is a measure of surface. For example, to measure the size of a room, you determine the area of the floor.

In the U.S. customary system, the most common area units are the square inch (sq in.), square foot (sq ft), and square yard (sq yd). In the metric system, the most common area units are the square centimeter (cm^2) and the square meter (m^2).

The area of a square is given by the formula $A = s^2$, where s stands for *side.*

- To find the area of a square, multiply the side by itself.

Square

$A = 3 \times 3$
$= 9$ area units

The area of a rectangle is given by the formula $A = lw$, where l stands for *length* and w stands for *width.*

- To find the area of a rectangle, multiply the length by the width.

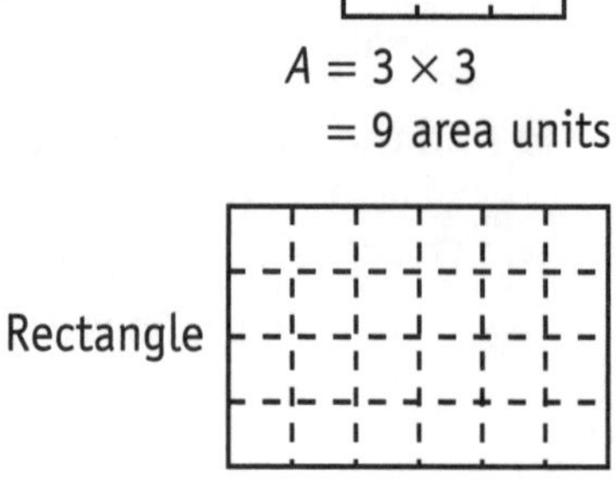

$A = 6 \times 4$
$= 24$ area units

EXAMPLE 1 What is the area of a square that measures 5 feet on each side?

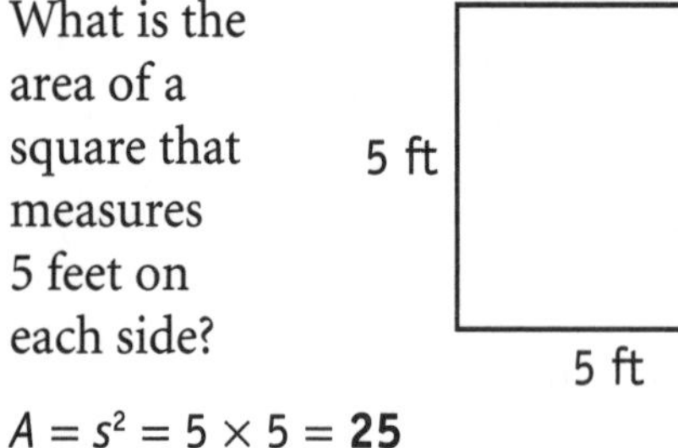

$A = s^2 = 5 \times 5 = \mathbf{25}$

ANSWER: 25 sq ft

EXAMPLE 2 What is the area of a rectangle that is 12 inches long and 6 inches wide?

6 in.
12 in.

$A = l \times w = 12 \times 6 = \mathbf{72}$

ANSWER: 72 sq in.

Find the area of each figure.

1. $A =$ ____________ **2.** $A =$ ____________ **3.** $A =$ ____________

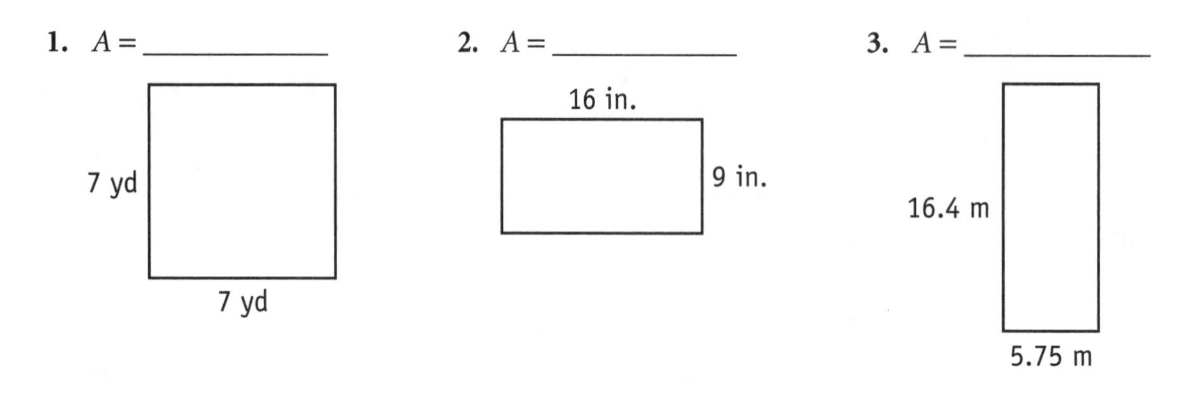

Area of a Triangle

The area of a triangle is given by the formula $A = \frac{1}{2}bh$, where b represents the **base** and h represents the **height.**

The base is one side of the triangle. The height is the distance from the base to the vertex of the opposite side. Only in a right triangle is the height one of the sides of the triangle.

To find the area of a triangle, multiply $\frac{1}{2}$ times the base times the height.

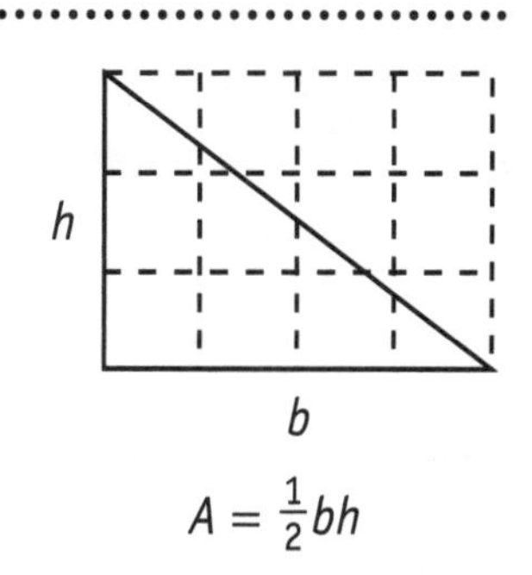

$A = \frac{1}{2}bh$

EXAMPLE 1

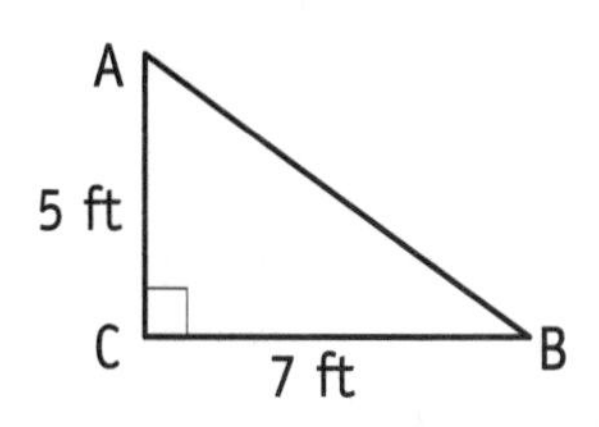

The height (5) is a side of right triangle ABC.

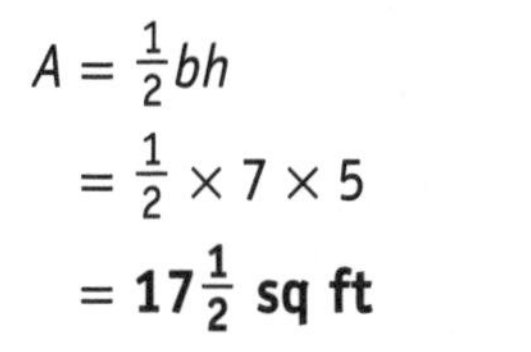

$A = \frac{1}{2}bh$

$= \frac{1}{2} \times 7 \times 5$

$= \mathbf{17\frac{1}{2}}$ **sq ft**

EXAMPLE 2

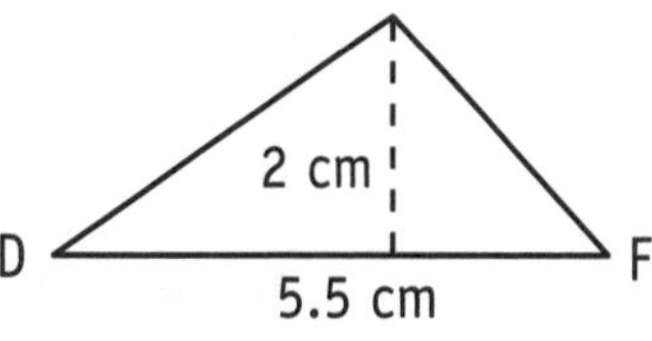

The height (2) is drawn as a dotted line within ΔDEF.

$A = \frac{1}{2}bh$

$= \frac{1}{2} \times 5.5 \times 2$

$=$ **5.5 sq cm**

EXAMPLE 3

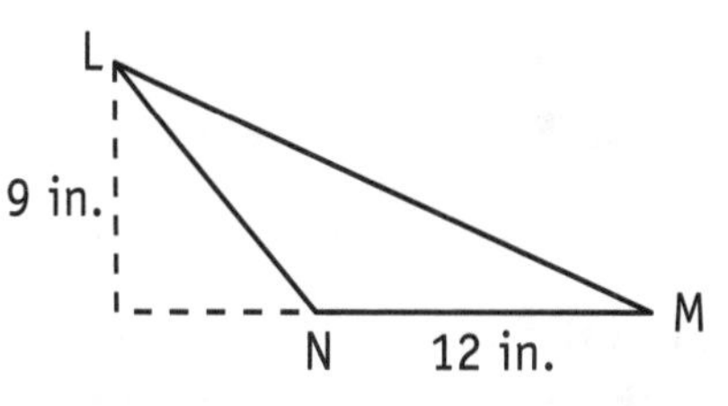

The height (9) is drawn as a dotted line outside of ΔLMN.

$A = \frac{1}{2}bh$

$= \frac{1}{2} \times 12 \times 9$

$=$ **54 sq in.**

Find the area of each triangle.

1. $A =$ ____________ **2.** $A =$ ____________ **3.** $A =$ ____________

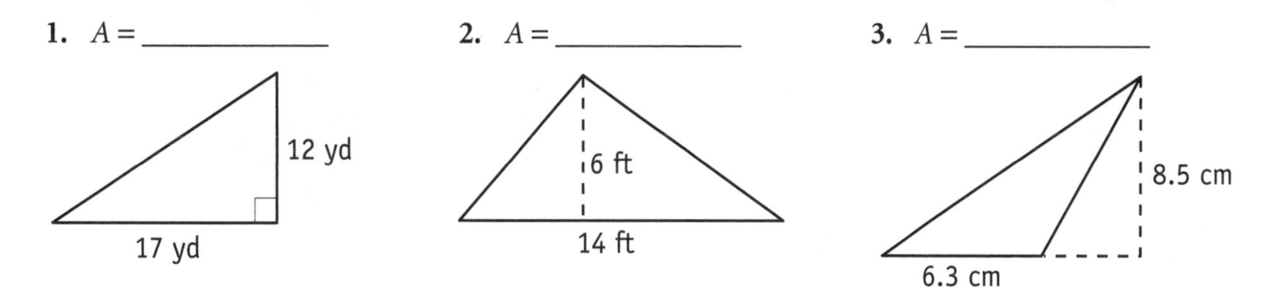

4. A piece of cloth, in the shape of an equilateral triangle, is 4 feet long on each side and 4.5 feet high. What is the area of this piece of cloth?

5. A piece of plywood is in the shape of a right triangle. How many square feet is the plywood if the base is 4 feet wide and the height is 8 feet?

Area of a Circle

The area of a circle is given by the formula $A = \pi r^2$, where $\pi = 3.14$ or $\pi = \frac{22}{7}$, and $r =$ radius.

To find the area of a circle, multiply π times the radius times the radius.

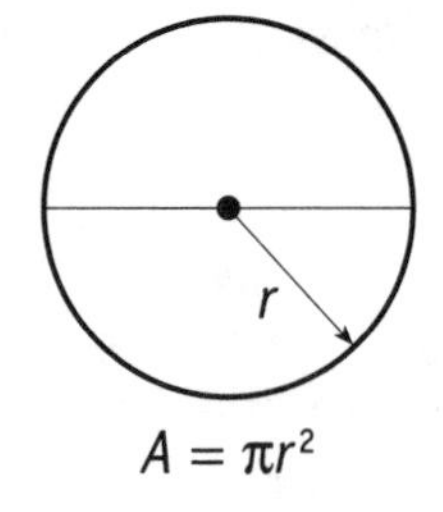

$A = \pi r^2$

EXAMPLE A circular garden has a radius of 21 feet. Find the area of the garden.

Write $\frac{22}{7}$ for π, and 21 for r in the area formula.

$A = \pi \times r \times r = \frac{22}{\cancel{7}_1} \times \overset{3}{\cancel{21}} \times 21 = \mathbf{1,386}$

ANSWER: 1,386 square feet

Find the approximate area of each circle. Round each decimal answer to the nearest whole number.

1. $A =$ ____________

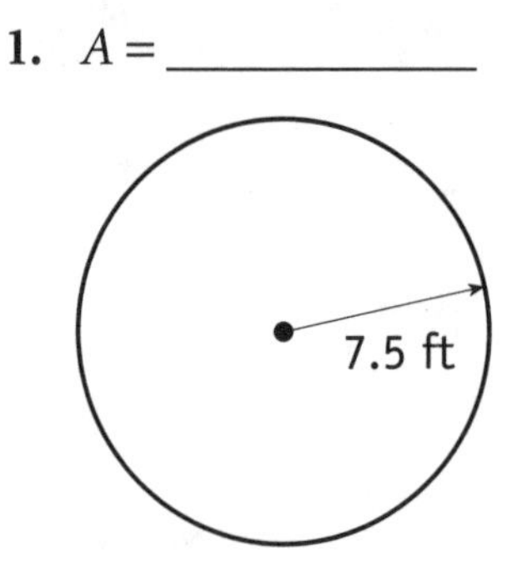

2. $A =$ ____________

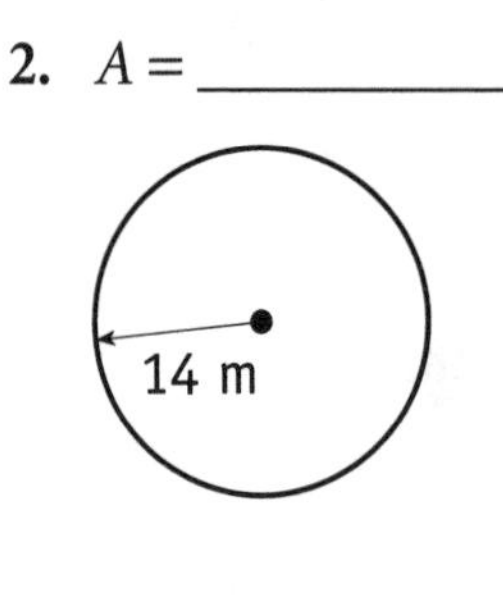

3. $A =$ ____________

18 in.

4. A circular wading pool has a radius of 7 feet. What is the area of the surface of the pool?

5. A sprinkler sprays water in a circular pattern. The spray reaches 4.9 meters from the sprinkler head. *Estimate* the lawn area watered by the sprinkler.

6. The circular top of a library table has a diameter of 6.5 feet. To the nearest square foot, what is the area of the tabletop?

7. To the nearest 0.1 square foot, what is the area of the window pictured at the right if $r = 2$ feet?

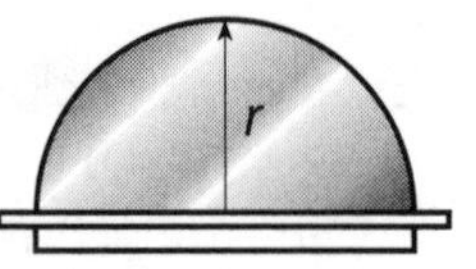

Reviewing Perimeter, Circumference, and Area

For problems 1–3, find the distance around each figure.

1. $P =$ ___________ 2. $P =$ ___________ 3. $C =$ ___________

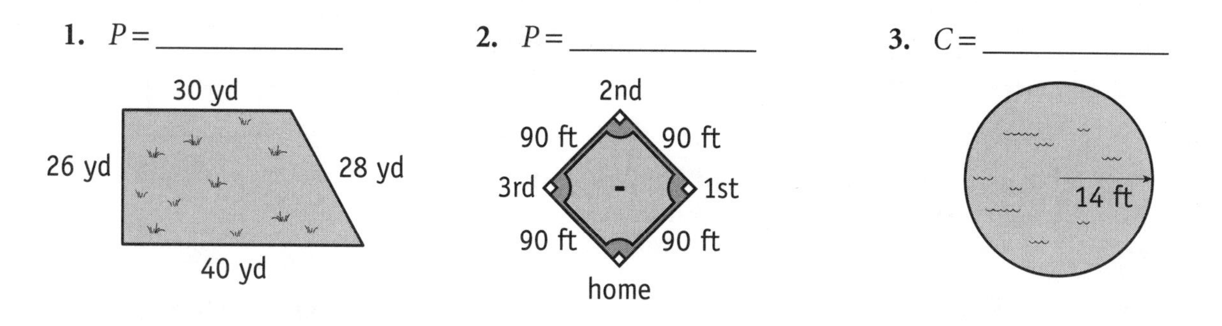

For problems 4–6, find the area of each figure.

4. $A =$ ___________ 5. $A =$ ___________ 6. $A =$ ___________

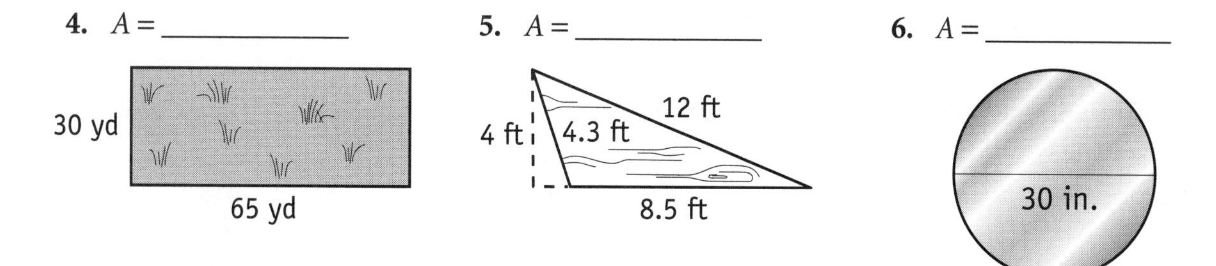

Solve each problem.

7. A concrete slab that is 20 feet long and 14 feet wide is to be covered in fancy square tiles.

 a. If each tile measures 1 foot along each side, how many tiles will be needed?

 b. If each tile measures 8 inches along each side, about how many tiles will be needed?

8. Mai Lin designed the prop for a school play shown at the right. She wants to find the area of the surface shown.

 a. What is the area of the square bottom section?

 b. What is the approximate area of the top half-circle section?

 c. What is the total area of the surface?

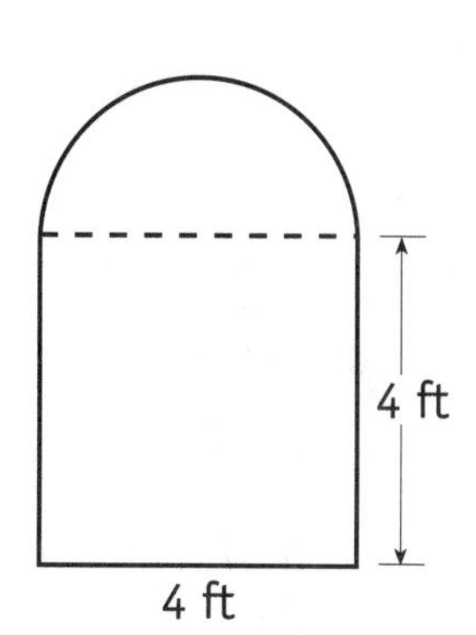

Volume of a Rectangular Solid

Volume (V) is a measure of the space taken up by a solid object, such as a brick, or is a measure of the space enclosed in a freezer, box, or other.

The volume of a rectangular solid is given by the formula $V = l \times w \times h$, where l = length, w = width, and h = height.

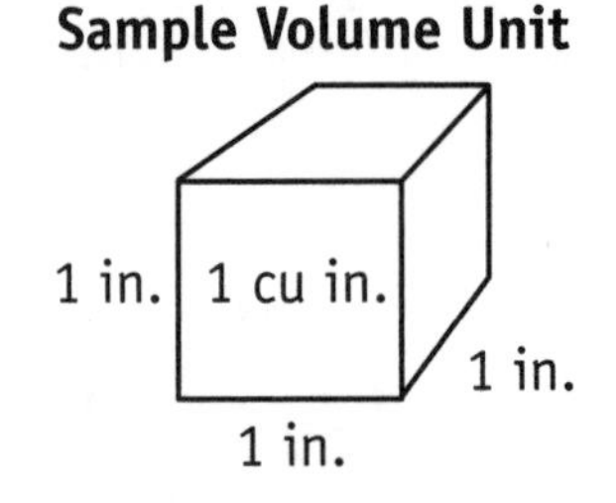

In the U.S. customary system, common volume units are the cubic inch (cu in.), cubic foot (cu ft) and cubic yard (cu yd). In the metric system, common volume units are the cubic centimeter (cm^3) and the cubic meter (m^3).

Most often, you'll be asked to find the shape of a **rectangular solid**—the math name for familiar shapes such as boxes, suitcases, freezers, and rooms. A rectangular solid is also called a **rectangular prism.**

A rectangular solid with equal-length sides is called a **cube.**

To find the volume of a retangular solid, multiply the length times the width times the height.

$$V = l \times w \times h$$

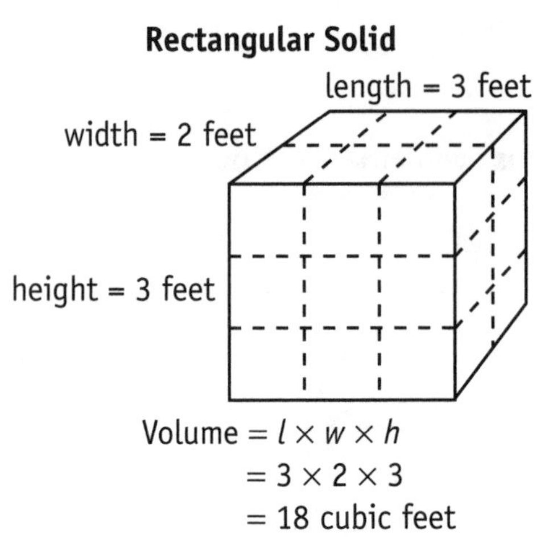

Find the volume of each rectangular solid. If necessary, round your answers to the nearest whole number.

1. ____________ 2. ____________ 3. ____________

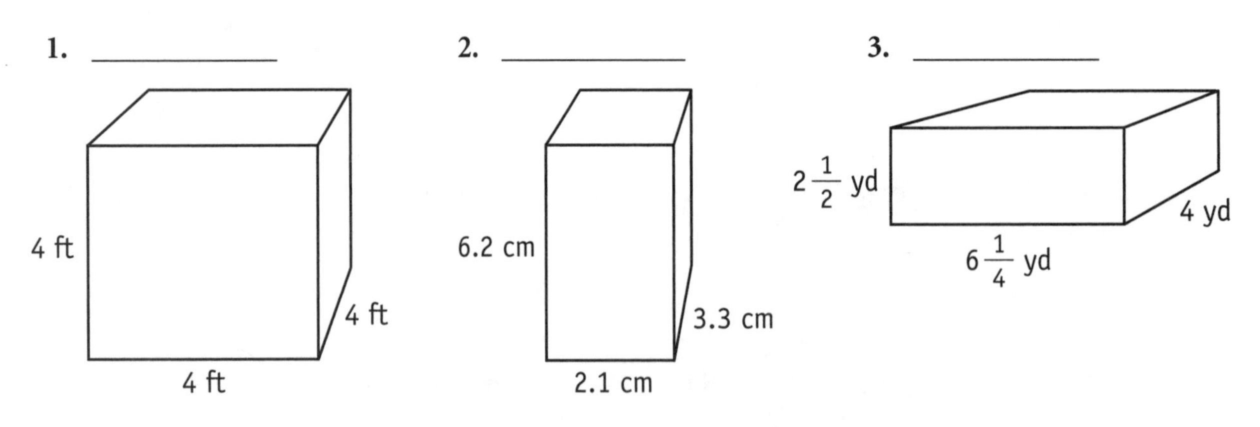

4. An aquarium in the shape of a rectangular solid measures 3 feet by 2 feet by 1.5 feet.
   a. What is the volume of the aquarium?
   b. About how many gallons of water does the aquarium hold? (1 cubic foot ≈ 7.5 gallons)

5. A waterbed measures 6 feet long, 5 feet wide, and 6 inches high. If water weighs about 62 pounds per cubic foot, what is the approximate weight of the waterbed when full?

Volume of a Cylinder

The volume of a cylinder is given by the formula

$V = \pi r^2 h$, where $\pi \approx 3.14$ or $\frac{22}{7}$, r = radius, and h = height.

To find the volume of a cylinder multiply the area of the bottom (a circle with area πr^2) times the height.

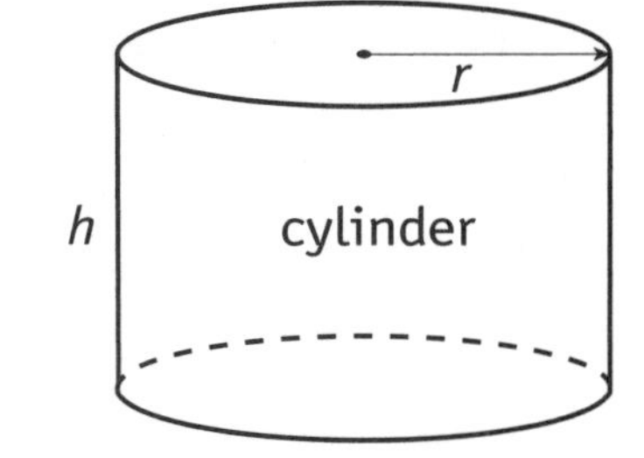

The top and bottom of a cylinder are equal-size circles.

EXAMPLE A cylindrical oil drum has a radius of 2 feet and a height of 5 feet. About how many cubic feet of oil does this drum hold when full?

$$V = \pi r^2 h = \pi \times r \times r \times h \approx 3.14 \times 2 \times 2 \times 5$$
$$\approx 3.14 \times 20$$
$$\approx 62.8$$

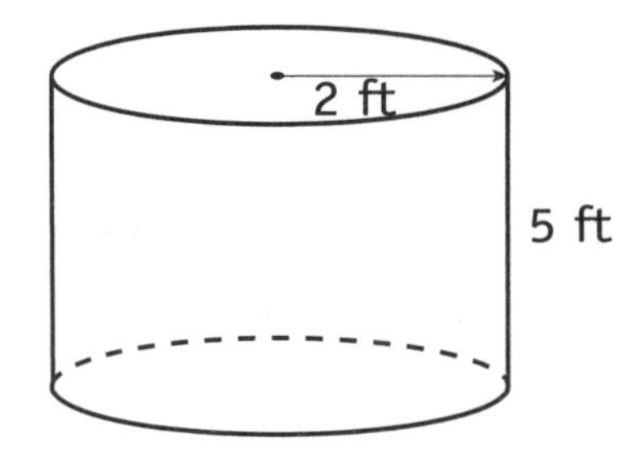

ANSWER: about 63 cubic feet

Find the volume of each cylinder below. If necessary, round your answers to the nearest tenth.

1. ____________

2. ____________

3. ____________

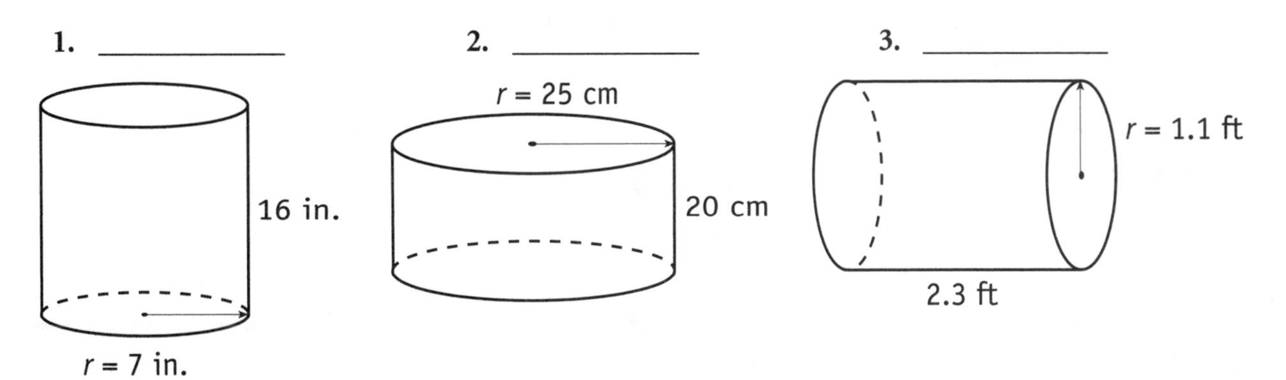

4. A cylindrical oil drum is 5 feet tall and has a diameter of 1.5 feet.

   a. To the nearest cubic foot, how many cubic feet of oil can the drum hold?

   b. If the drum is two-thirds full, what volume of oil does it now hold?

5. A cylindrical water tower is 18 feet tall and has a diameter of 12 feet.

   a. To the nearest cubic foot, about how many cubic feet of water can the tower hold?

   b. About how many gallons can the tower hold? (1 cubic foot $\approx$ 7.5 gallons)

Geometry and Measurement Review

Work each problem and check your answers. Correct any errors.

1. Draw a line that is parallel to line A.

2. Draw a line that is perpendicular to line B.

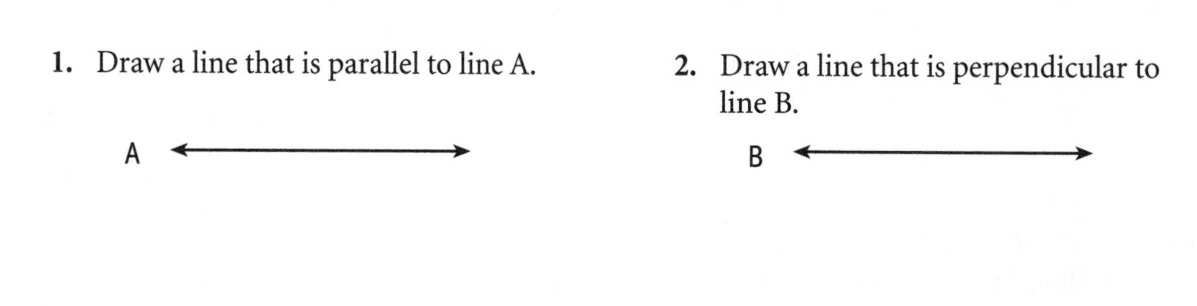

3. Draw an angle that measures 90°.

4. Draw an angle that measures 180°.

Name each angle below. Then classify each angle as acute, right, obtuse, straight, or reflex.

5.

6.

7.

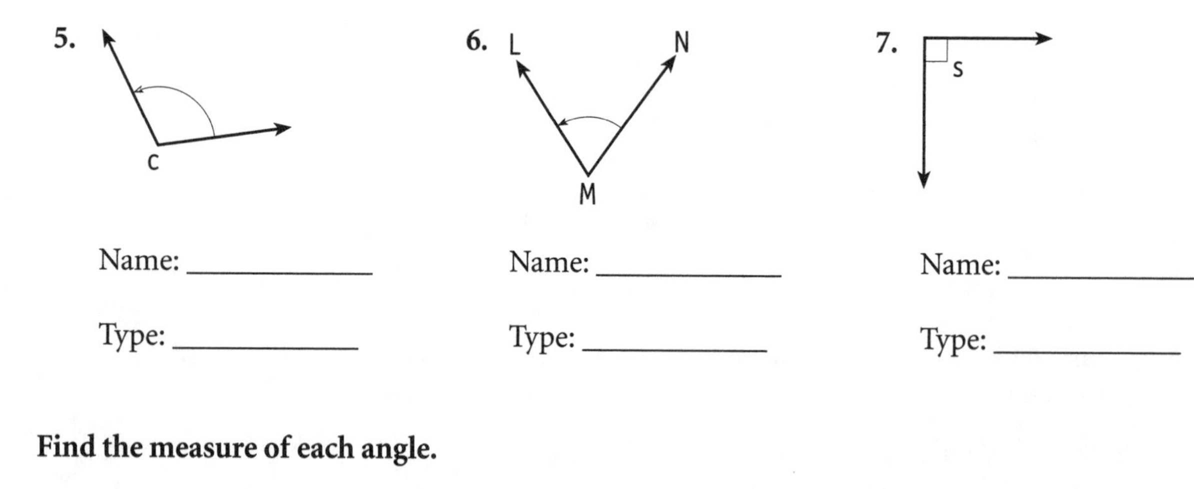

Name: ____________ Name: ____________ Name: ____________

Type: ____________ Type: ____________ Type: ____________

Find the measure of each angle.

8. ∠GEF = ______

9. ∠ACD = ______

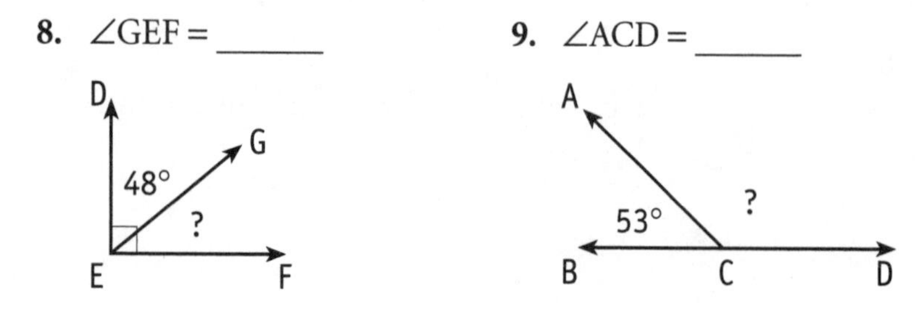

10. ∠x = ______

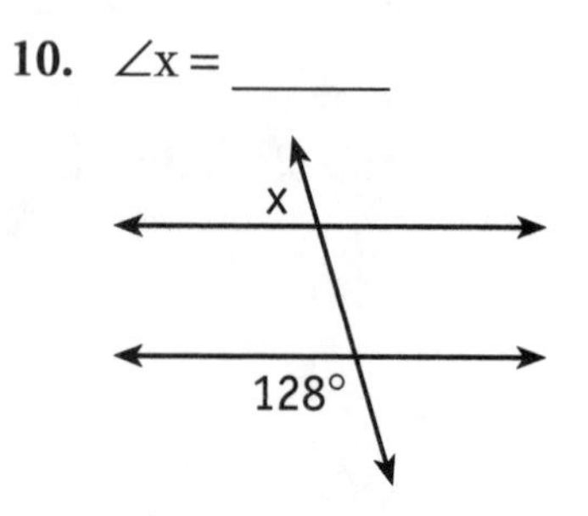

11. What type of triangle is shown below: equilateral, isosceles, or right?

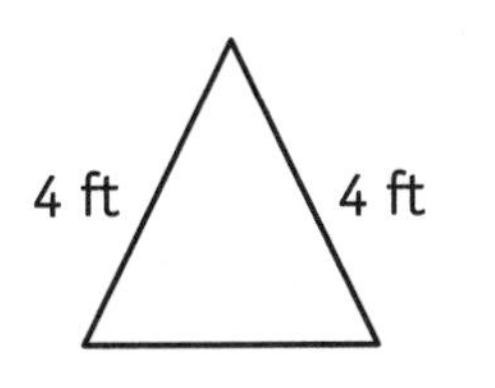

12. What is the measure of ∠B?

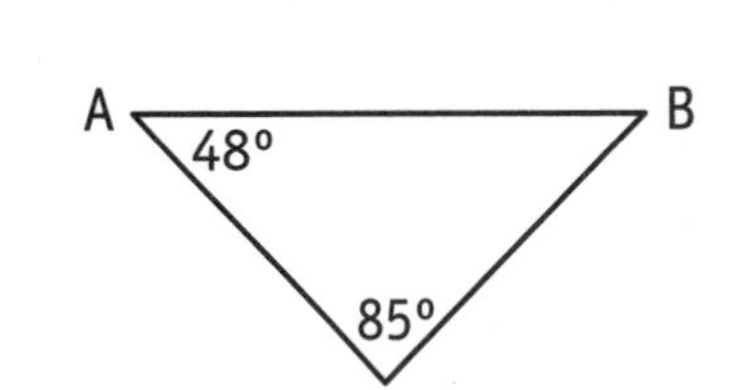

For problems 13–17, refer to the drawing at the right.

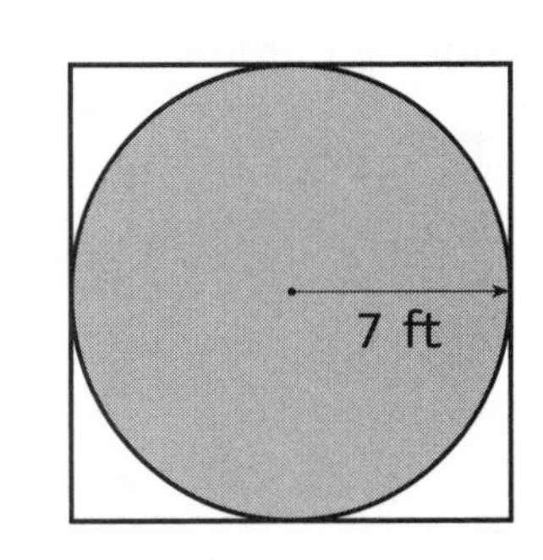

13. What is the perimeter of the square?

14. Estimate the circumference of the circle.

15. What is the area of the square?

16. Estimate the area of the circle.

17. Estimate the ratio of the area of the circle to the area of the square.

18. Find the area of ΔABC.

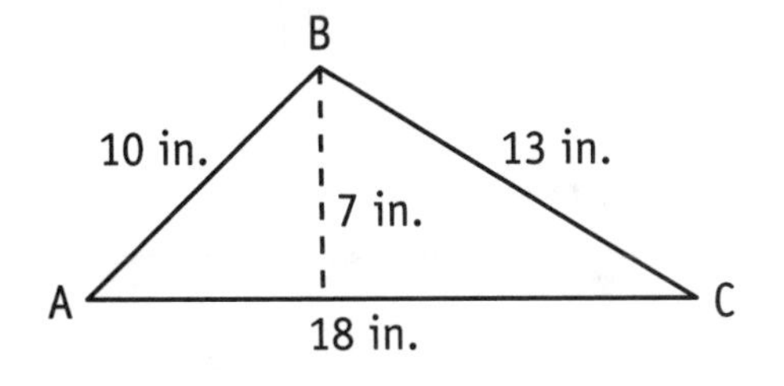

19. Find the volume of the rectangular solid.

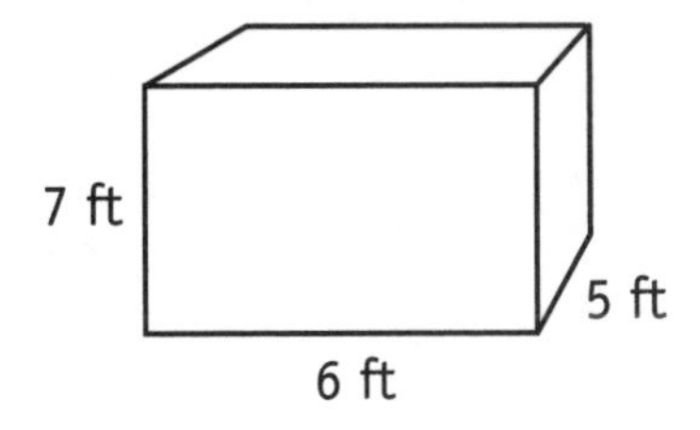

20. To the nearest cubic foot, what is the volume of the cylinder? ($\pi \approx 3.14$)

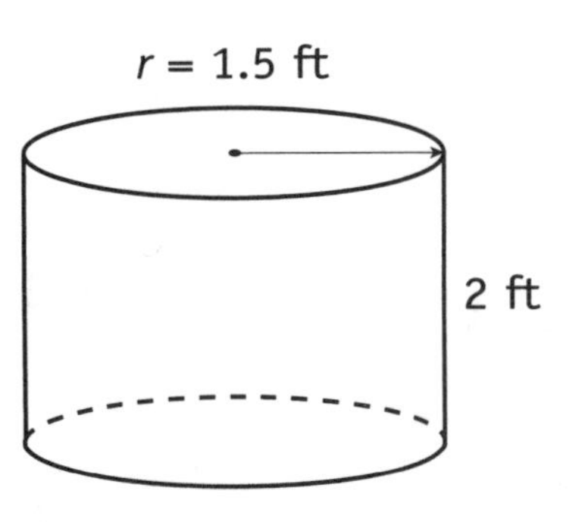

SPATIAL SENSE AND PATTERNS

Symmetry

A figure has a **line of symmetry** if you can divide the figure into two halves that are mirror images of each other. The figure is said to have symmetry around the line that divides it.

- A line of symmetry is drawn through the heart below.
- The central vein is a line of symmetry for the leaf.
- The piece of wood has no line of symmetry.

Some figures have more than one line of symmetry.

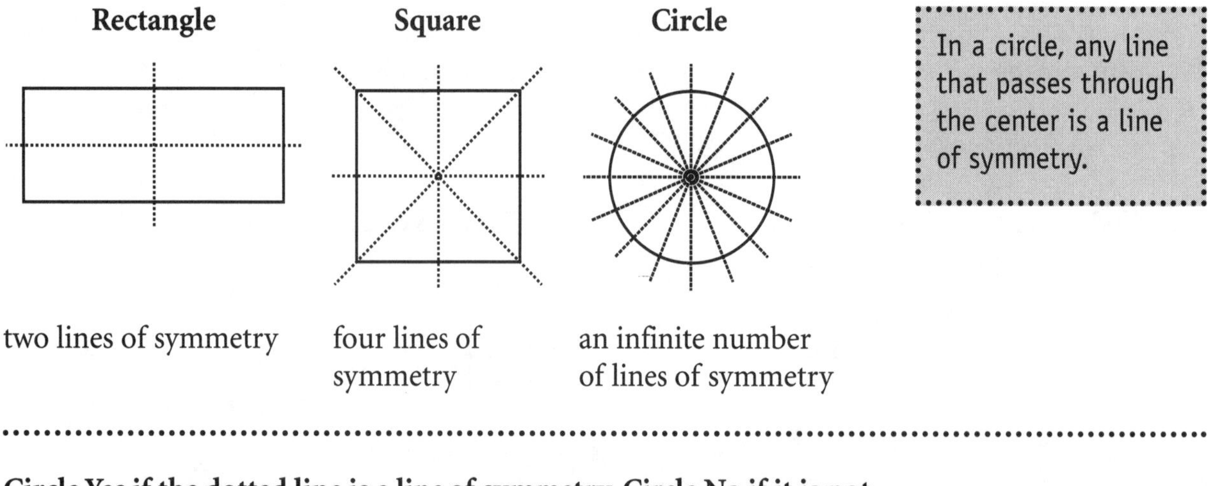

In a circle, any line that passes through the center is a line of symmetry.

Circle Yes if the dotted line is a line of symmetry. Circle No if it is not.

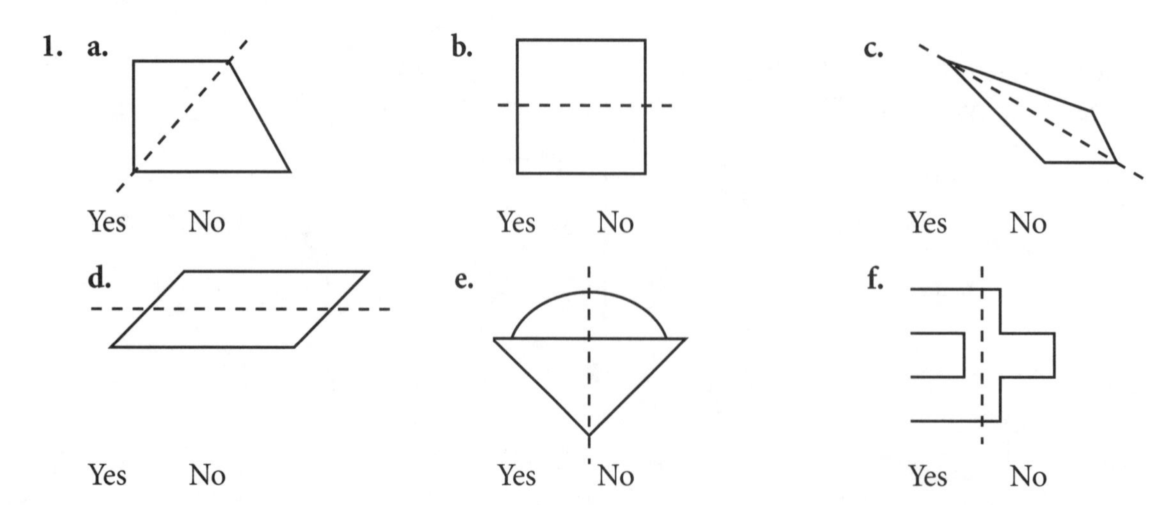

Draw one line of symmetry in each figure below.

2. a. **b.** **c.**

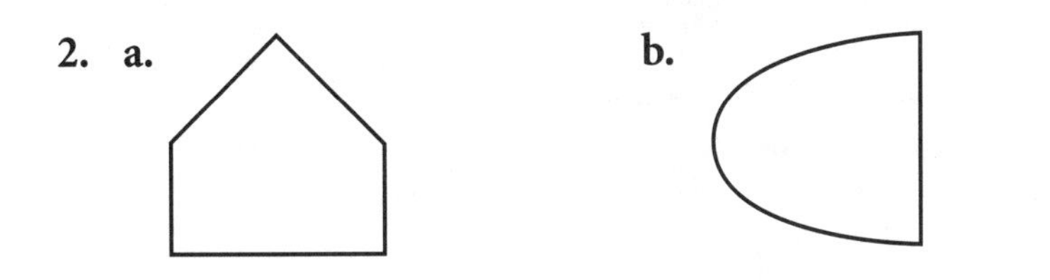

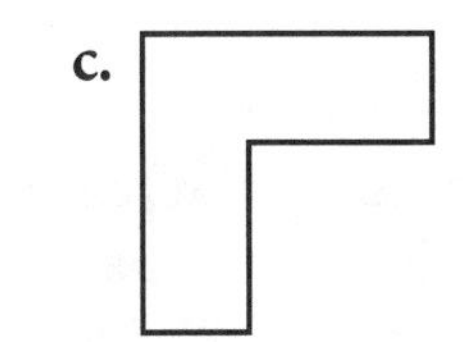

Draw all possible lines of symmetry in each figure below.

3. a. **b.** **c.**

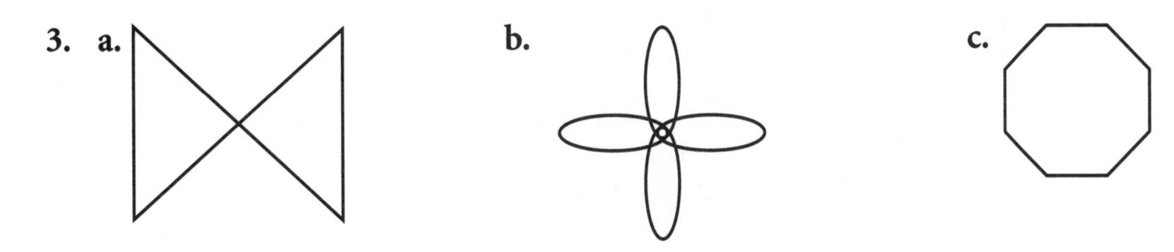

Half of a figure is drawn on each grid below. Only the half to the left of a line of symmetry is shown. Draw the other half of each figure.

4. a. **b.**

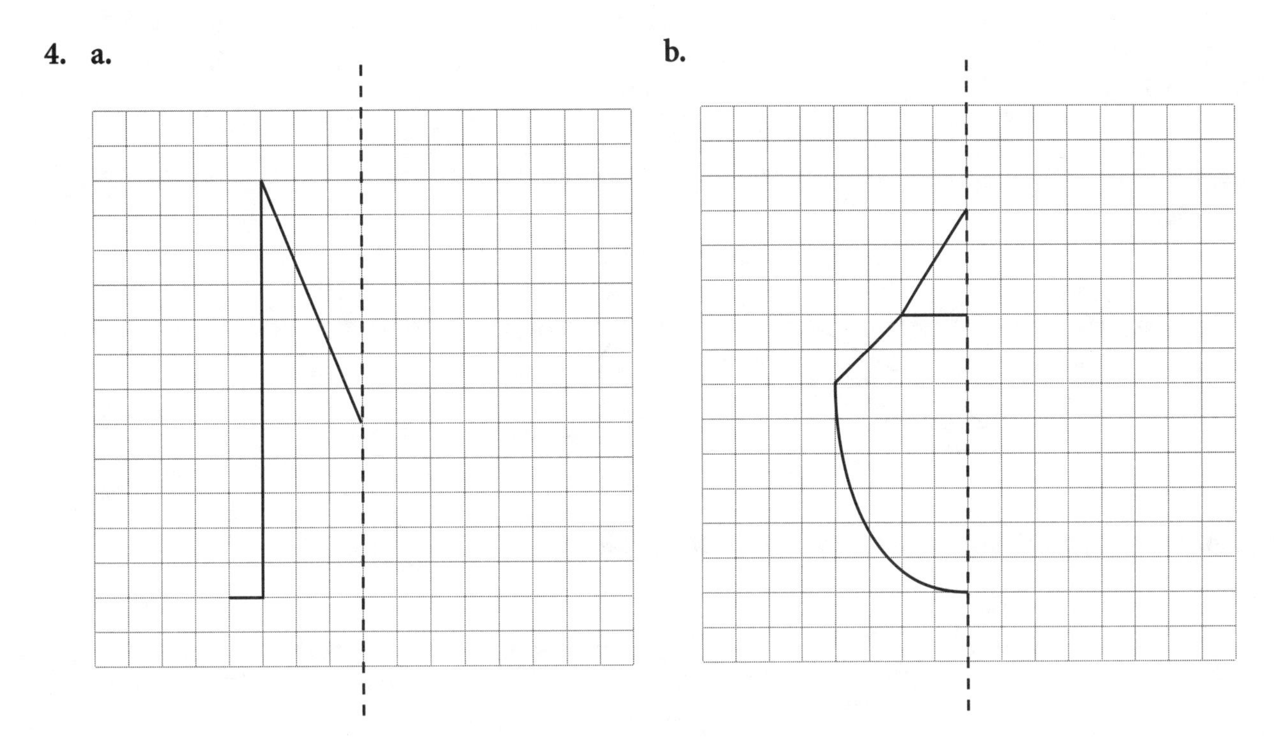

Congruent Figures and Similar Figures

Two or more figures that look alike may be congruent or similar.

- **Congruent figures** are exactly the same size and shape.
- **Similar figures** have exactly the same shape but may not be the same size. *Figures that are congruent are also similar.*

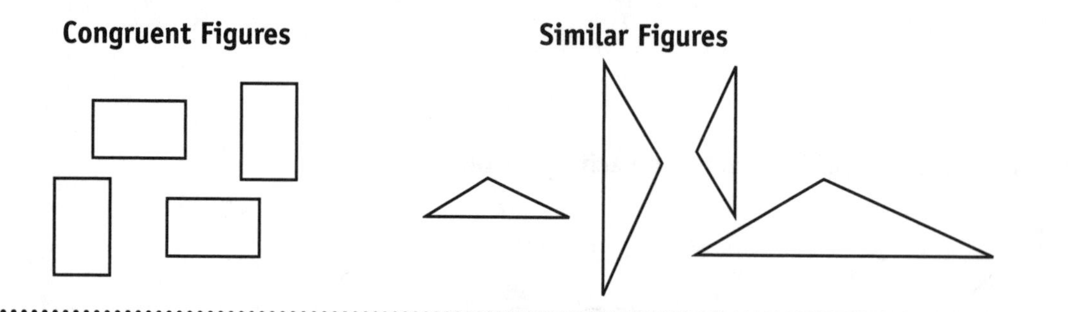

For problems 1–3, circle *two* figures that are similar to the first figure.

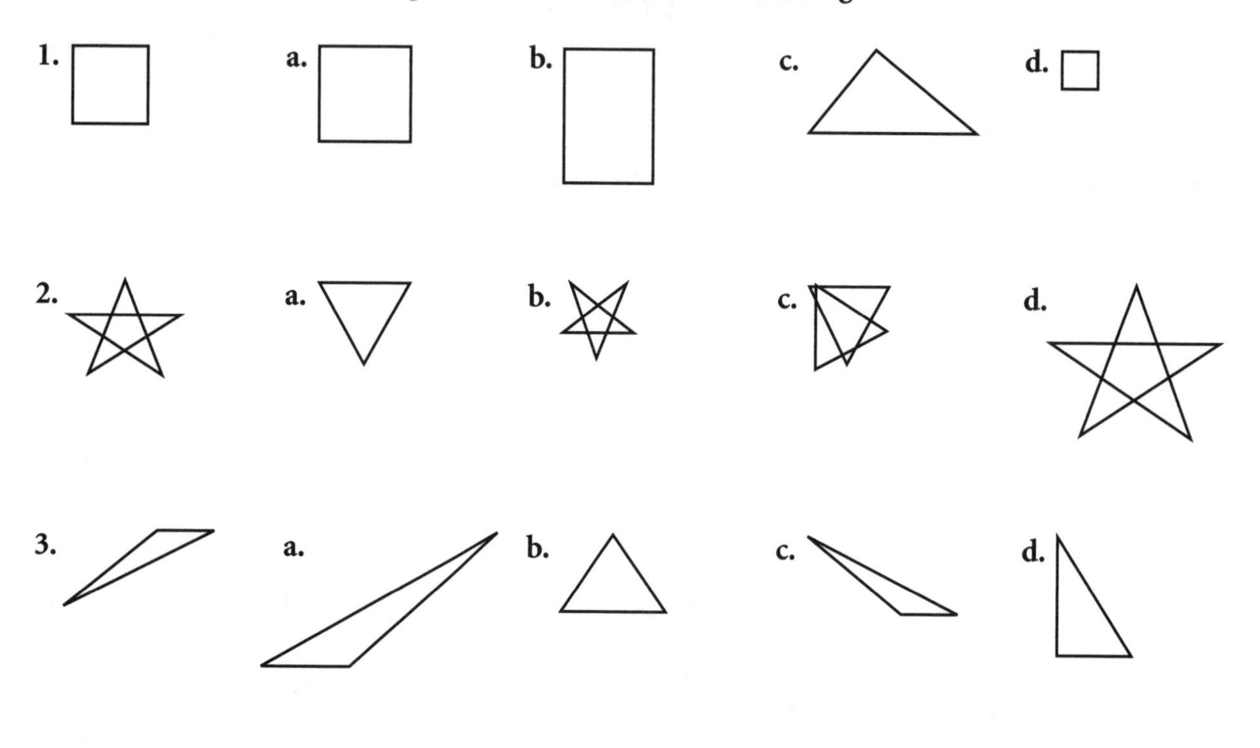

4. When you enlarge a photograph, you are making a ________________ figure.
(congruent *or* similar)

5. Draw two triangles that are congruent.

6. Draw two triangles that are similar but not congruent.

Tessellations

A **tessellation** is a pattern formed by drawing congruent figures side by side, without leaving gaps and without overlapping. A honeycomb is a tessellation. In a honeycomb, the repeating figure is a **regular hexagon,** a six-sided figure with equal sides and equal angles.

Tessellations are used in the design of floor tiles, quilts, rugs, curtains, and many other things.

Many figures such as squares, rectangles, and regular hexagons can be used to make tessellations. Circles cannot. Circles cannot be placed together side by side without creating gaps.

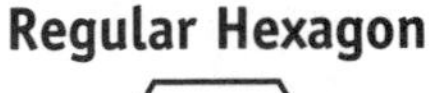

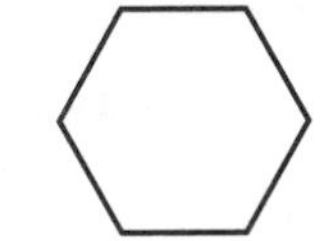

Honeycomb Tessellation

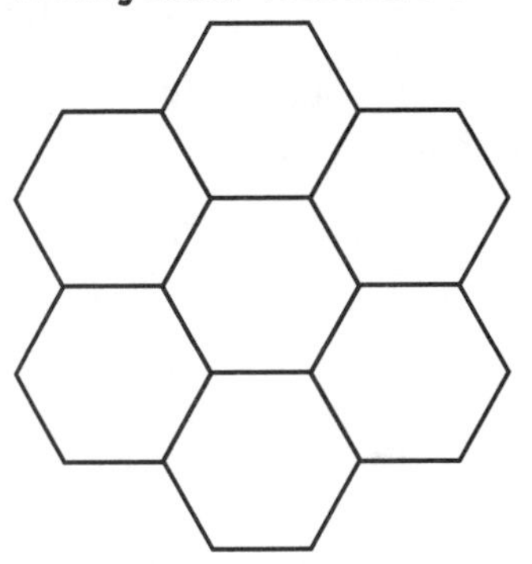

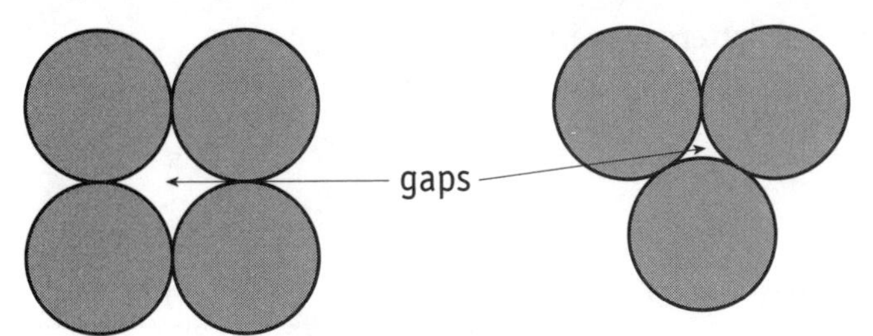

Circles cannot be used to form a tessellation.
There is no way to get rid of the gaps!

Three figures are drawn below.

- **On scratch paper draw several copies of each figure.**
- **Cut out the figures and see if any can be used to form a tessellation.**
- **Make a pattern of each figure, seeing if you can fill the space between them. You can turn the figures in any direction.**

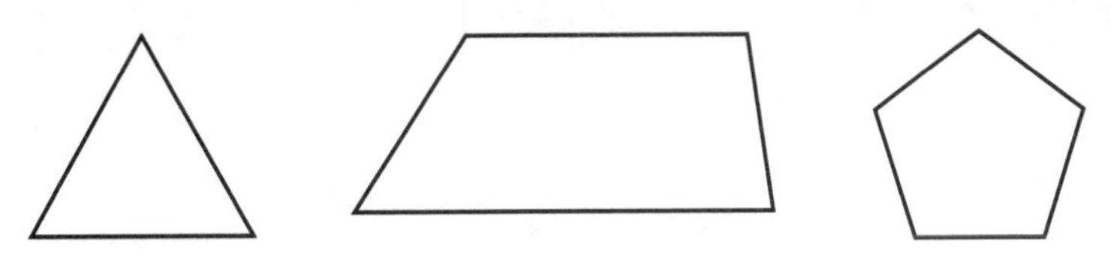

1. Circle each figure that can be used for a tessellation.

2. For each figure that you circled in problem 1, draw its tessellation pattern below.

Using Map Coordinates

Directions tell us where things are—how things are related in space. On a map, both map directions (*north, south, east,* and *west*) and map coordinates may be used.

Map directions tell us the direction one location is from another.

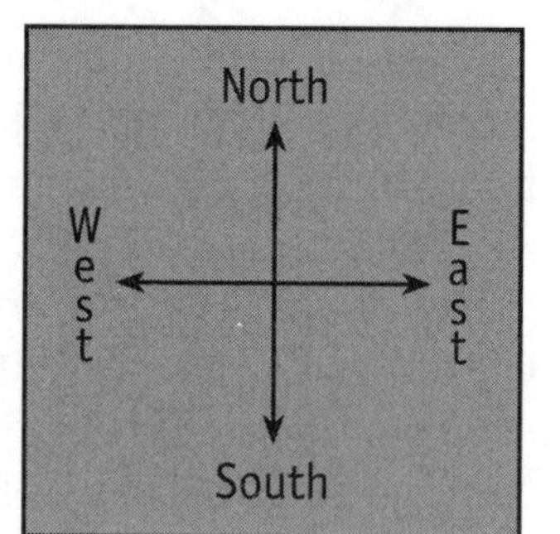

- On most maps, north is located at the top, south at the bottom, west at the left, and east at the right.

Map coordinates are often used on city maps to help us find a particular location on the map itself.

- **Coordinates** are one or more numbers or letters that identify a region or a single point on a map.

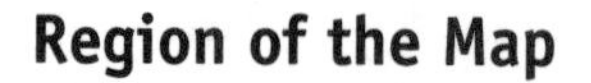

Region of the Map

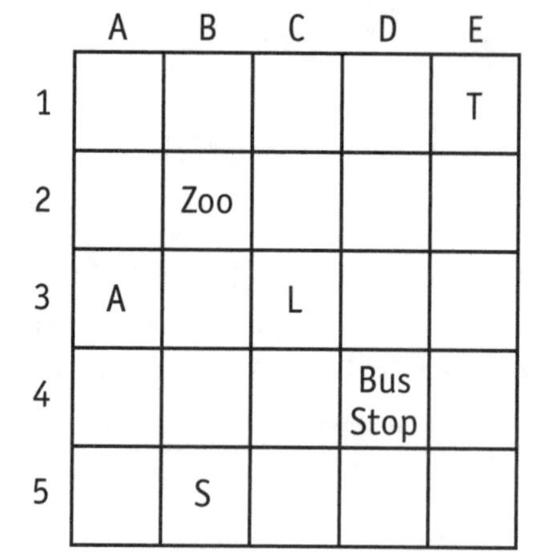

Map A is divided into labeled squares. Each square makes up a region of the map. *Each square is identified by the letter directly above it and by the number directly to its left.*

EXAMPLE 1 Suppose you want to identify the square in which the zoo is located and the square in which the bus stop is located. To do so, give the coordinates of each square as a letter followed by a number.

- The zoo is located in square **B2.**
- The bus stop is located in square **D4.**

Intersection of Streets

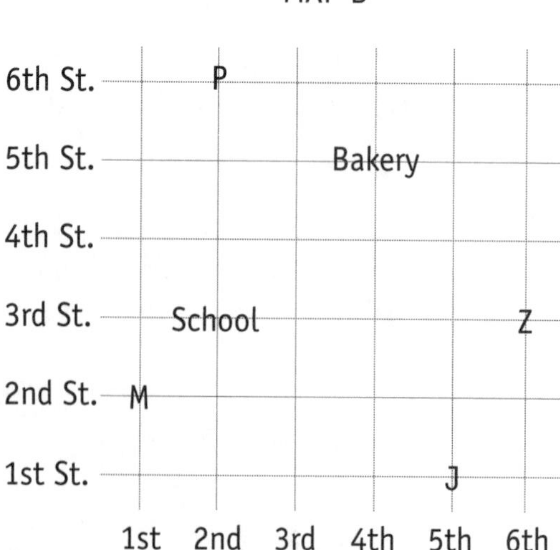

On Map B, *coordinates are given as the names of two streets.* The coordinates identify the point at which the two streets intersect (cross). The name of the street running north and south (up and down) is usually given first.

EXAMPLE 2 Suppose you want to know the locations of the school and the bakery.

- The school is located at (or near) the intersection of **2nd Ave** and **3rd Street.**
- The bakery is located at (or near) the intersection of **4th Ave** and **5th Street.**

For problems 1–3, refer to Map A on page 128. Assume north is at the top of Map A.

1. Write the coordinates of the square in which each is found.

EXAMPLE **Zoo** B2

a. Theater (**T**) ______
b. Library (**L**) ______
c. Anne's house (**A**) ______
d. Swimming pool (**S**) ______

2. To get from the bus stop to the zoo, you should walk

______ blocks ____________ and ______ blocks ____________.
number direction number direction

3. Tilley's Restaurant is located three blocks east and one block south of the zoo. What are the coordinates of the square in which Tilley's Restaurant is found?

For problems 4–6, refer to Map B on page 128. Assume north is at the top of Map B.

4. Write the coordinates of the intersection at which each place is located.

EXAMPLE **School** 2nd Ave., 3rd St.

a. Pizza restaurant (**P**) ______ Ave., ______ St.
b. Zoo (**Z**) ______ Ave., ______ St.
c. Maria's house (**M**) ______ Ave., ______ St.
d. Jake's Market (**J**) ______ Ave., ______ St.

5. To get from the bakery to the school, you should walk

______ blocks ____________ and ______ blocks ____________.
number direction number direction

6. The cycle shop is located two blocks west and four blocks south of the bakery. What are the coordinates of the intersection at which the cycle shop is located?

Becoming Familiar with a Coordinate Grid

A **coordinate grid** is formed by combining a vertical number line with a horizontal number line.

- The vertical number line is called the ***y*-axis.**
- The horizontal number line is called the ***x*-axis.**
- The point at which the two lines meet is called the **origin.**

Writing Coordinates

Every point on a coordinate grid has coordinates, two numbers that tell its position.

- The ***x*-coordinate** (*x*-value) tells how far the point is from the *y*-axis.
  Positive *x* indicates the point is to the right of the *y*-axis.
  Negative *x* indicates the point is to the left of the *y*-axis.
- The ***y*-coordinate** (*y*-value) tells how far the point is from the *x*-axis.
  Positive *y* indicates the point is above the *x*-axis.
  Negative *y* indicates the point is below the *x*-axis.
- Coordinates are usually written within parentheses, the *x*-coordinate followed by the *y*-coordinate.
- The origin has coordinates (0, 0).

EXAMPLE On the grid below, point A is (–6, 4).

x-coordinate ↑ ↑ *y*-coordinate

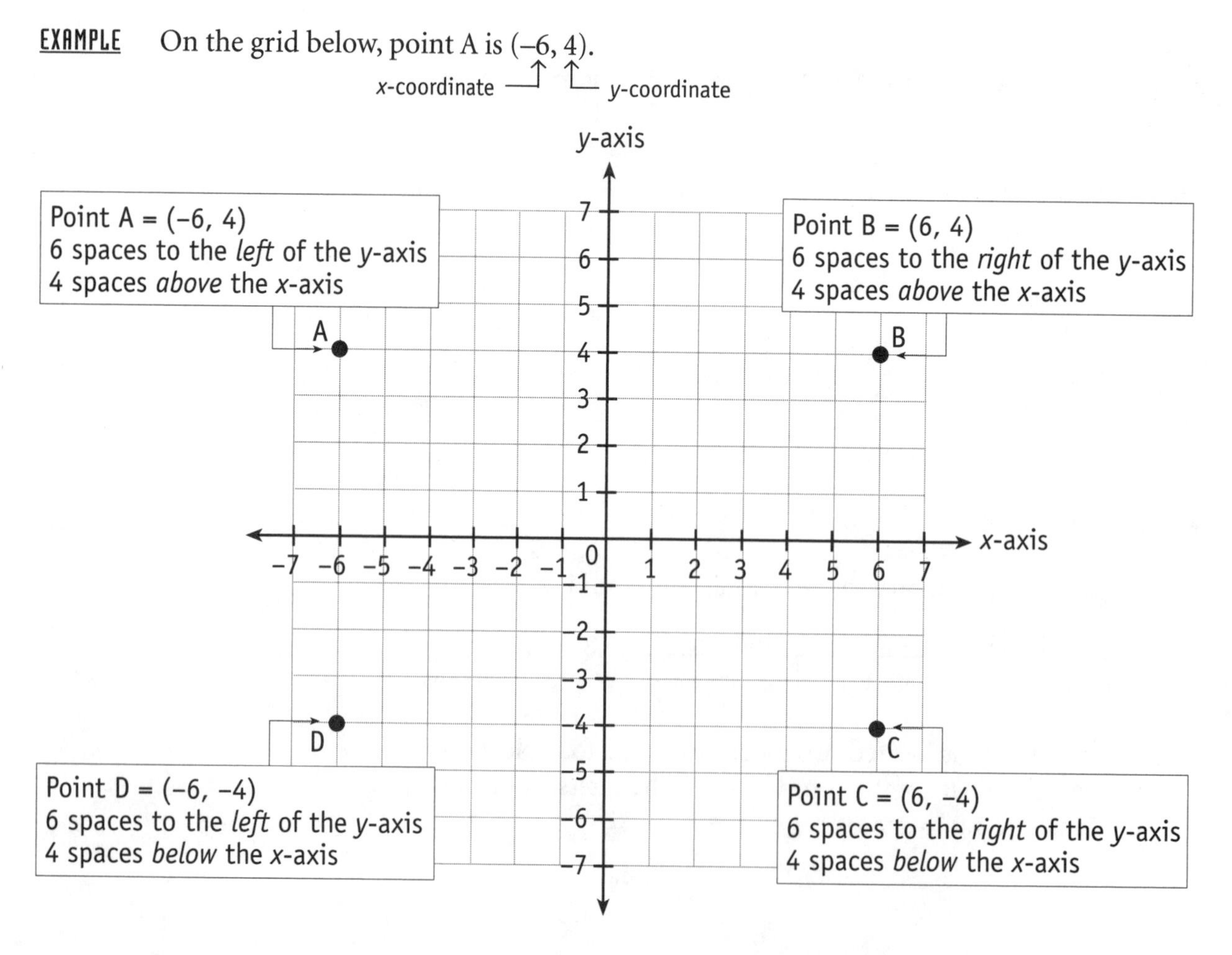

Write each pair of coordinates as numbers within parentheses. The first one is done as an example.

1. a. $x = 3, y = -5$ (3, −5) b. $x = -2, y = 6$ c. $x = 3, y = 1$ d. $x = -2, y = -7$

Identify the x- and y-coordinates of each point. The first one is done as an example.

2. a. Q = (2, 4)
   $x = 2$
   $y = 4$

   b. R = (−3, 5)
   $x =$
   $y =$

   c. S = (0, 4)
   $x =$
   $y =$

   d. T = (−5, 0)
   $x =$
   $y =$

Write the coordinates of each point shown on the grid. The first one is done as an example.

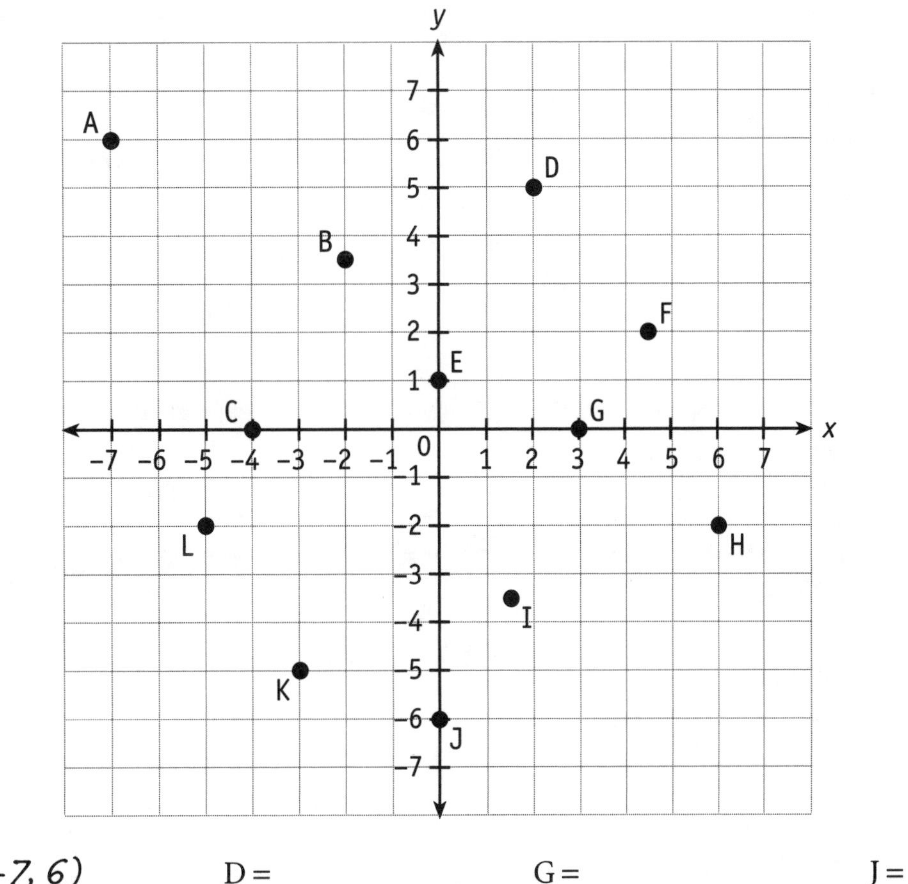

3. A = (−7, 6) D = G = J =

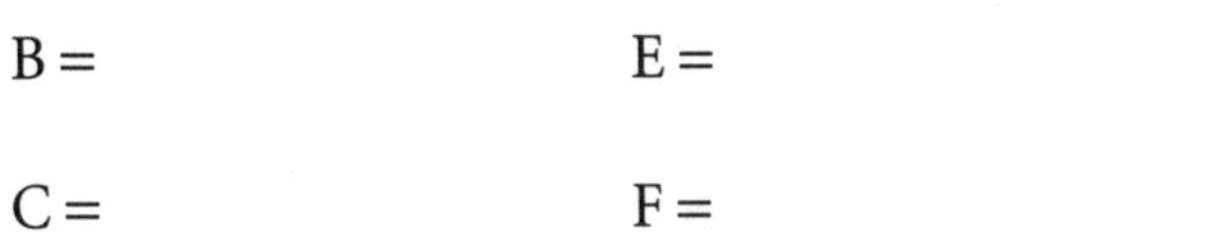

B = E = H = K =

C = F = I = L =

Working with Figures on a Coordinate Grid

Test questions often ask about figures drawn on a coordinate grid. On these next two pages are examples of the types of questions most often asked.

Completing a Figure

1. a. Points A, B, and C are the corner points of a square. Write the coordinates of each of these points.

   A = (,)

   B = (,)

   C = (,)

   b. Write the coordinates of point D, the fourth corner of the square.

   D = (,)

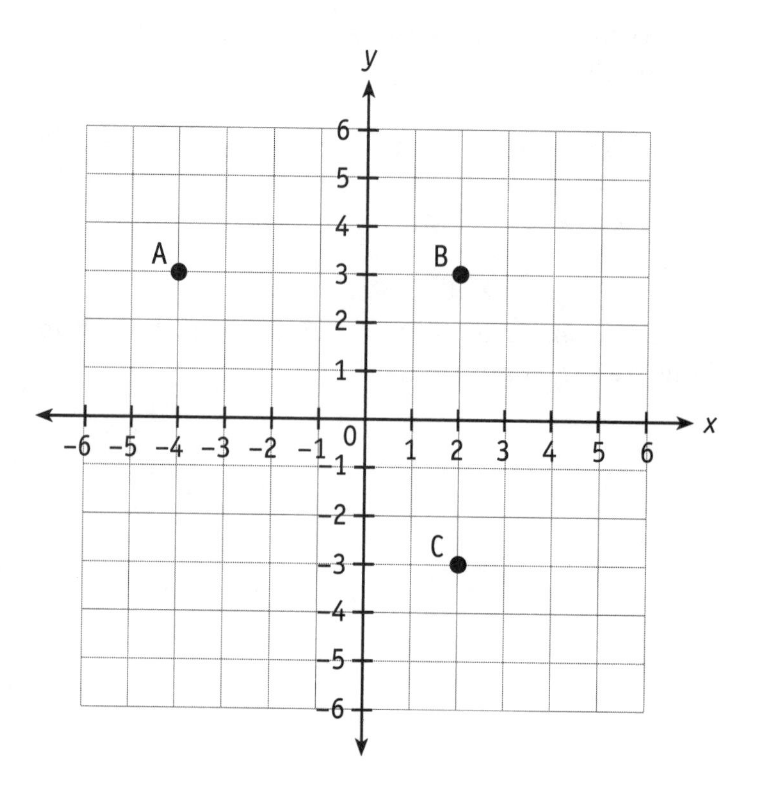

Identifying a Similar Figure

2. Rectangle EFGH is plotted on the grid at the right. Two coordinates for a second rectangle are also shown.

   Which of the following two additional coordinates complete the second rectangle so that the two rectangles are *similar*?

   a. (–3, 1) and (3, 1)

   b. (–3, –1) and (3, –1)

   c. (–3, –3) and (3, –3)

   d. (–3, –3) and (3, 5)

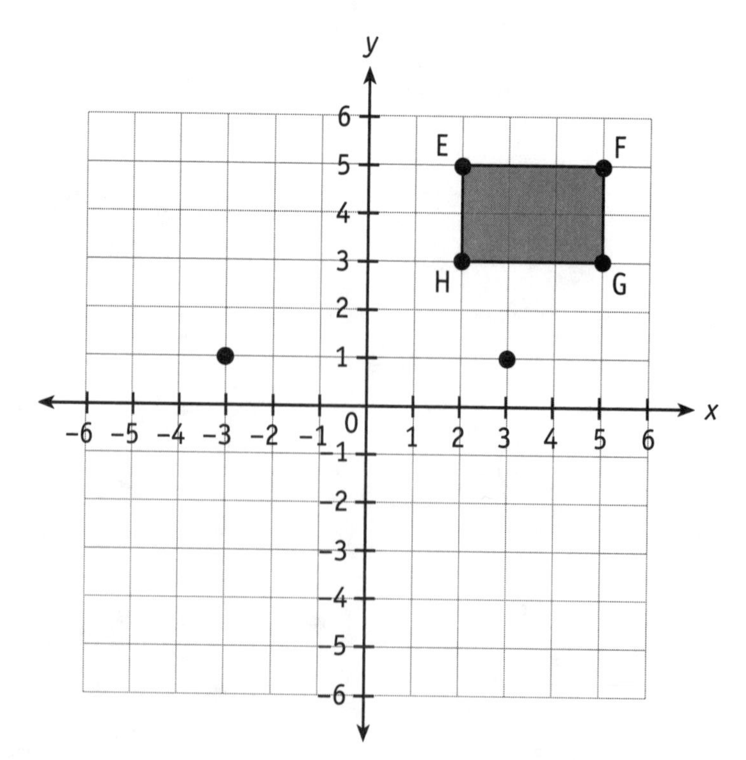

Sliding a Figure

- To *slide* is to move a figure without turning it.

3. Suppose you slide rectangle QRST 5 units to the right.

Write the coordinates of each vertex (corner) of the moved rectangle.

New Coordinates

Q = (,)

R = (,)

S = (,)

T = (,)

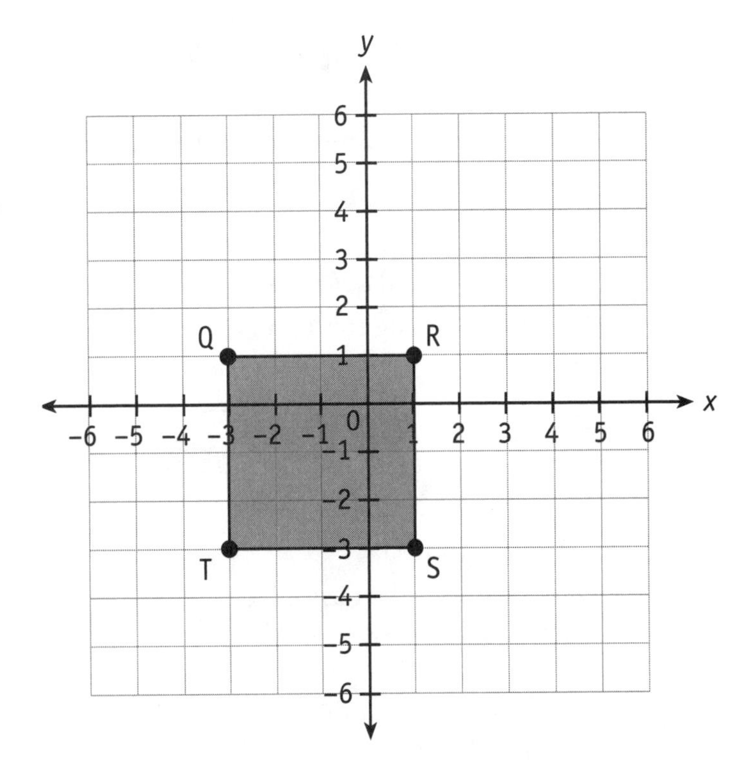

Rotating a Figure

- To *rotate* is to move a figure around a point.

4. Suppose you rotate triangle XYZ 90° clockwise (↷) around point Z.

Write the coordinates of each vertex (corner) of the rotated triangle.

New Coordinates

X = (,)

Y = (,)

Z = (,)

(Point Z does not change.)

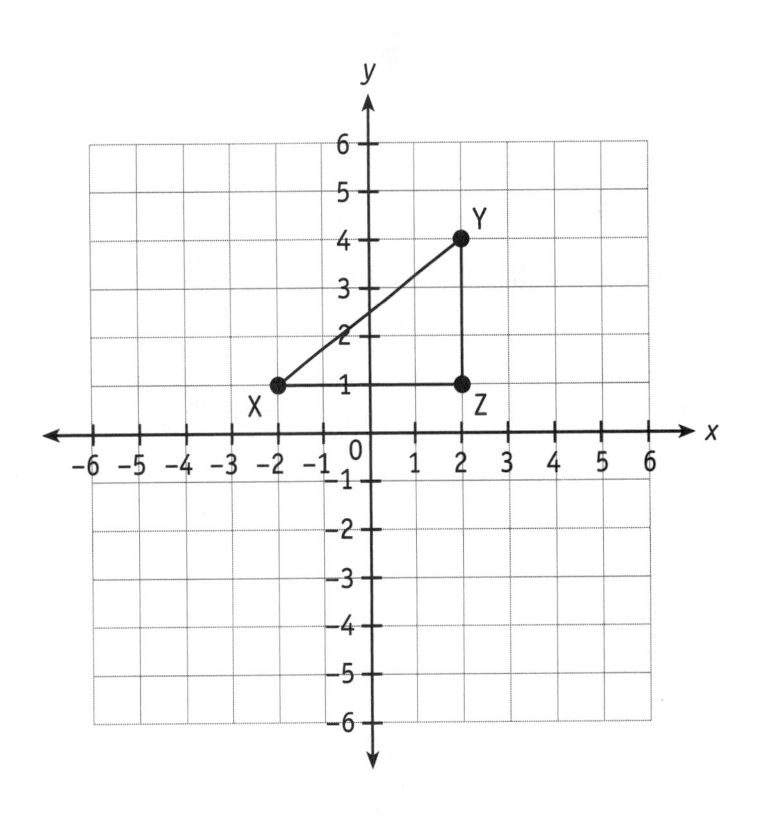

Patterns of Repeating Figures

A pattern of **repeating figures** is made up of a set of figures that repeat (continue the pattern by starting over).

EXAMPLE 1 The first three figures of a shaded pattern are shown below. What are the 4th and 5th figures in this pattern?

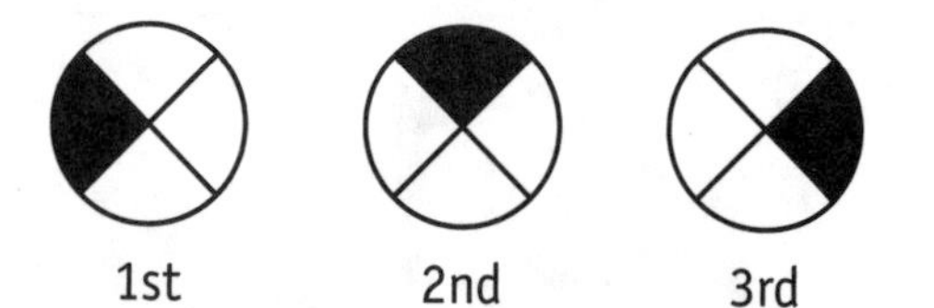

1st 2nd 3rd

In this pattern, the position of the shaded part moves one section clockwise with each new figure. The 4th figure continues this pattern. The 5th figure is the same as the first, and so on.

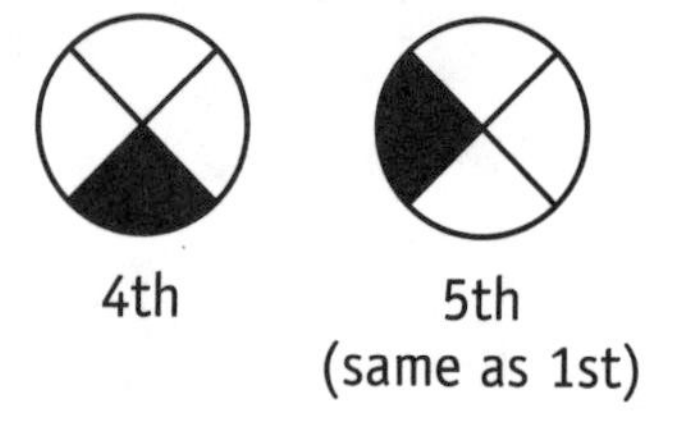

4th 5th (same as 1st)

EXAMPLE 2 The first five figures of a shape pattern are shown below. What are the 6th and 7th figures in this pattern?

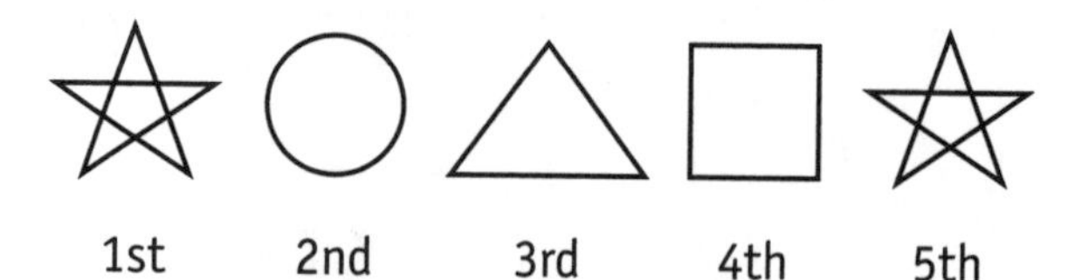

1st 2nd 3rd 4th 5th

This pattern is made up of four shapes that repeat. The 5th figure is a star, the same as the 1st figure. The 6th figure is a circle, the 7th is a triangle, and so on.

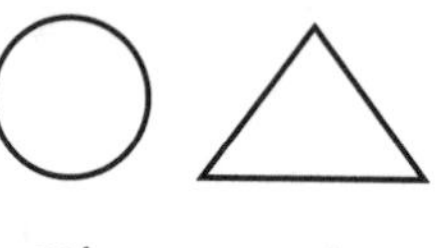

6th 7th

1. Complete the 4th and 5th figures of the shaded pattern below.

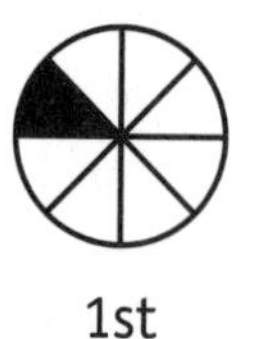
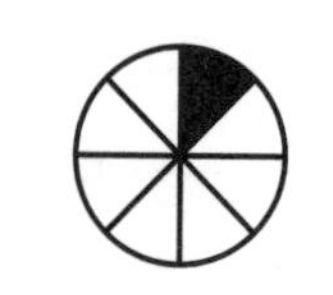

1st 2nd 3rd 4th 5th

2. Complete the 6th and 7th figures of the shaded pattern below.

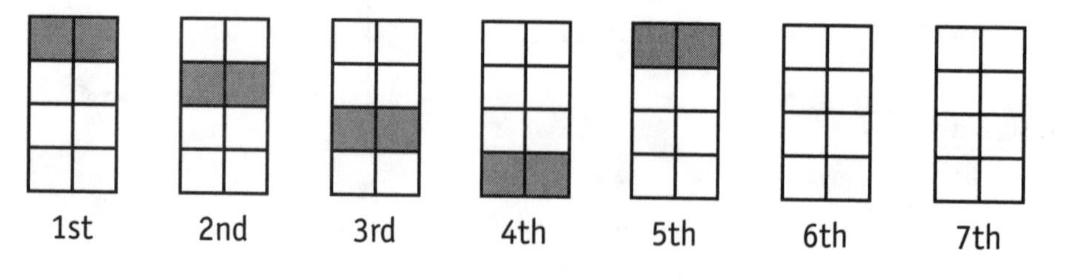

1st 2nd 3rd 4th 5th 6th 7th

3. Draw the 6th and 7th figures of the pattern below.

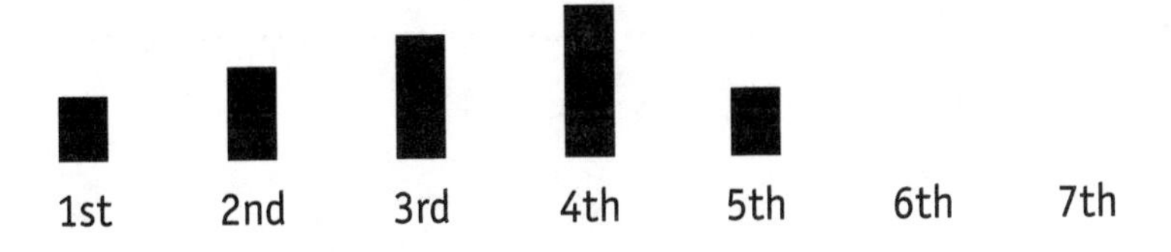

4. Draw the 4th and 5th figures in the angle pattern below.

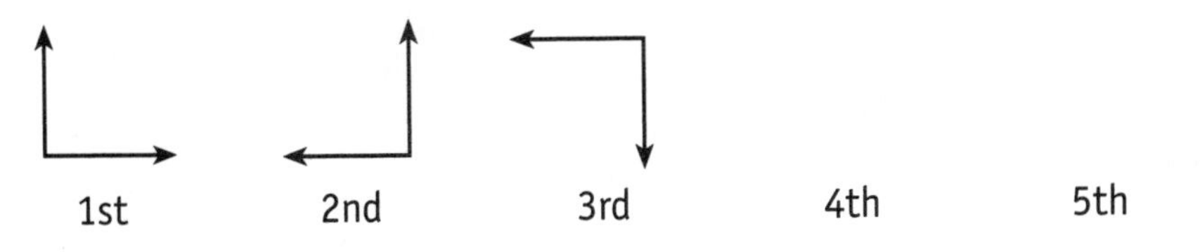

5. Draw the 7th and 8th figures in the picture pattern below.

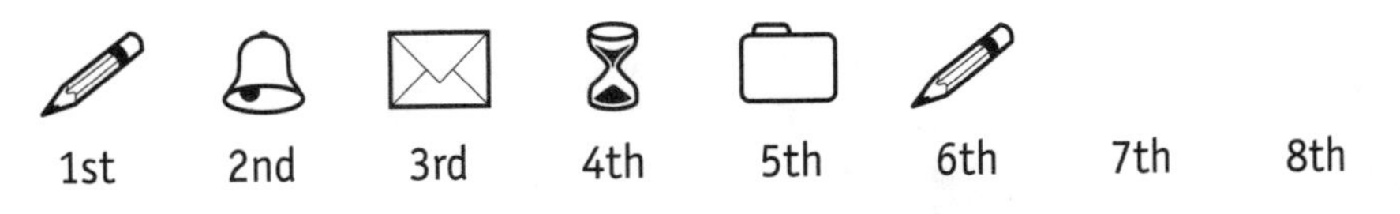

6. Look at the pattern of figures below.

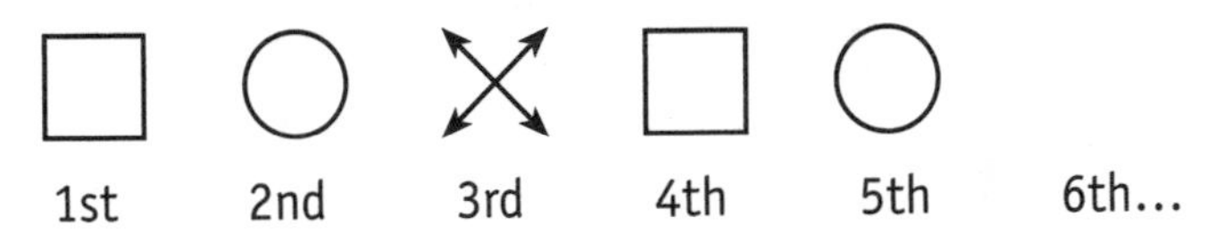

a. How many different shapes are shown? Do not count repeated shapes.

b. What is the 6th figure in the pattern?

c. What is the 12th figure in the pattern? (You do not need to draw 12 figures to determine the answer.)

d. What is the 100th figure in the pattern? (You do not need to draw 100 figures to determine the answer.)

Geometric Patterns

A **geometric pattern** is made up of a changing set of figures. Seeing the pattern enables you to determine other figures in the set.

EXAMPLE How many tiles are in the 4th and 5th figures of the tile pattern below?

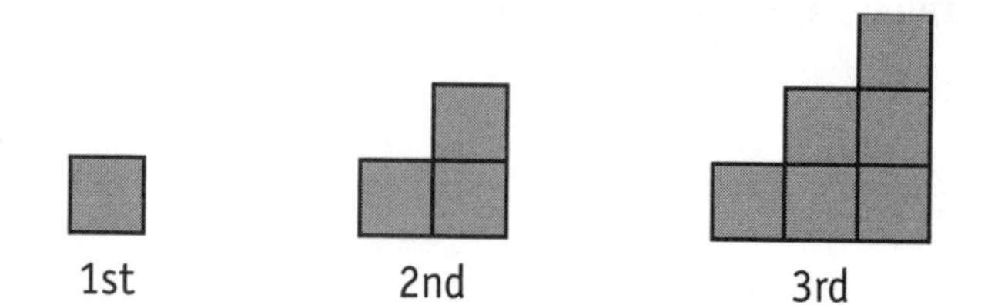

There are two ways to solve this problem.

Method I Draw the 4th and 5th figures and count tiles.

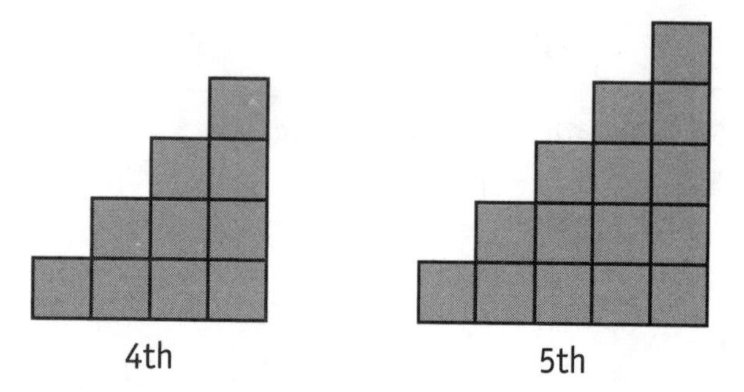

ANSWER: 4th figure = **10 tiles;** 5th figure = **15 tiles**

Method 2 Look for a pattern in the *number of tiles* in each figure.

Figure Number	**1**	**2**	**3**	**4**	**5**
Number of Tiles	1	3	6		

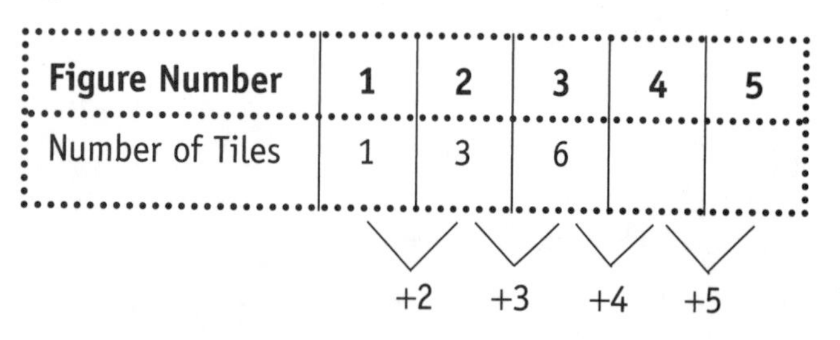

The number of tiles forms a number pattern. The 2nd figure has two more tiles than the 1st figure. The 3rd figure has three more tiles than the 2nd figure, and so on.

Continuing this pattern,

- the 4th figure has four more tiles than the 3rd figure.
- the 5th figure has five more tiles than the 4th figure.

ANSWER: 4th figure = 6 + 4 = **10 tiles**
5th figure = 10 + 5 = **15 tiles**

Solve each problem. Use any method you prefer.

1. The first three figures of a tile pattern are shown below.

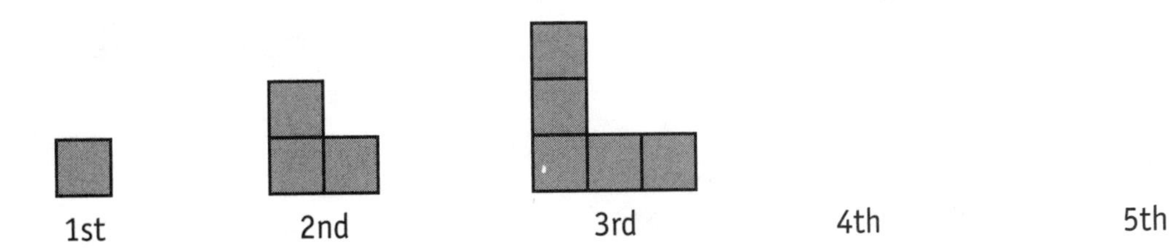

   a. Draw the 4th and 5th figures of this pattern.

   b. Write the first 10 numbers of the number pattern made by counting the *number of tiles* in each figure.

   ___ ___ ___ ___ ___ ___ ___ ___ ___ ___

   c. Write the first 10 numbers of the number pattern made by counting the total *number of right angles* in each figure. The first two numbers are given.

   4 13 ___ ___ ___ ___ ___ ___ ___ ___

2. A line segment is named by its endpoints. Three line segments—$\overline{XY}$, $\overline{XZ}$, and $\overline{YZ}$—are shown at the right.

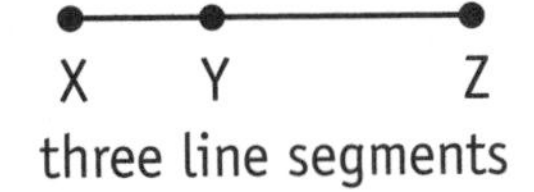

three line segments

   Below are the first three figures of a line-segment pattern.

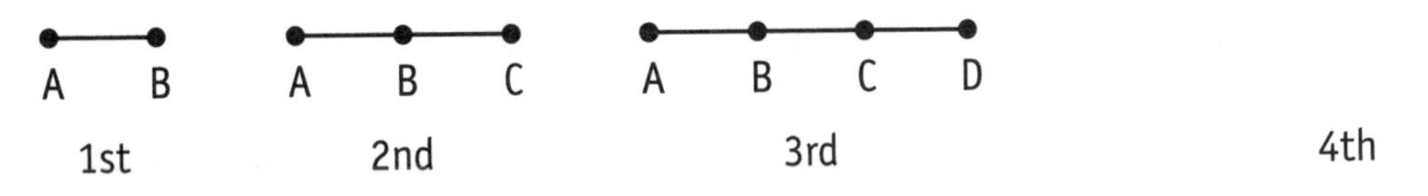

   a. Draw the 4th figure of this pattern.

   b. How many line segments are in each of the first four figures?

   1st:_____ 2nd:_____ 3rd:_____ 4th:_____

   c. The number of line segments in each figure is part of a number pattern. Write the first 10 numbers of this pattern (1st figure through 10th figure).

Figure Number	**1**	**2**	**3**	**4**	**5**	**6**	**7**	**8**	**9**	**10**
Number of Segments										

Spatial Sense and Patterns Review

Work each problem and check your answers. Correct any errors.

1. Draw all possible lines of symmetry in the oval.

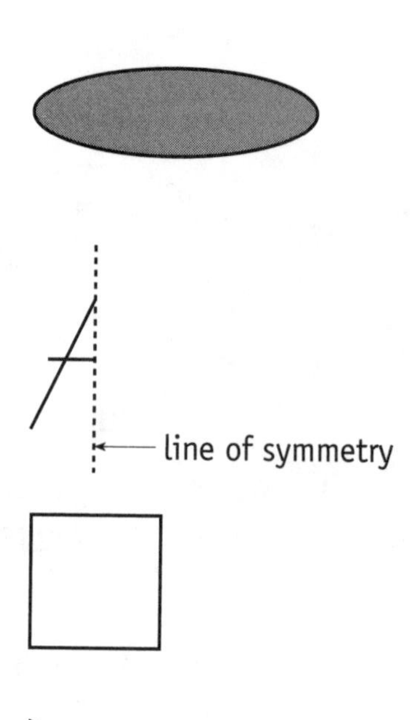

2. Half of a figure has been drawn on the left side of a line of symmetry. Complete the figure.

3. Draw a figure that is congruent to the square.

4. Draw a figure that is similar to the triangle and twice as wide.

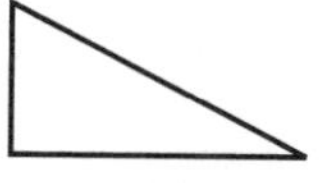

5. Draw two rectangles that are similar but not congruent.

6. Move the shaded square 5 squares east, 6 squares north, and 2 squares west. Shade the square of the new position.

N

7. Draw a tessellation using an isosceles triangle as the repeating figure.

8. What are the coordinates of the shaded square?

	A	B	C	D	E
1					
2					
3					
4					
5					

For problems 9 and 10, refer to the coordinate grid.

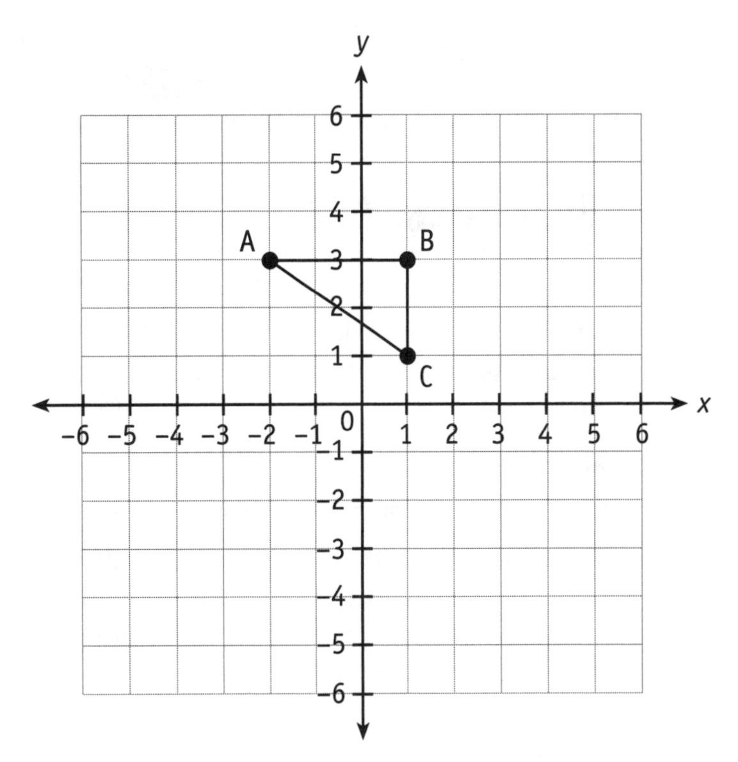

9. Write the coordinates of points A, B, and C.

A = (,)

B = (,)

C = (,)

10. Suppose you slide ΔABC 4 units to the right. What will be the new coordinates of points A, B, and C?

A = (,)

B = (,)

C = (,)

11. a. Shade the 4th and 5th figures in the shaded pattern below.

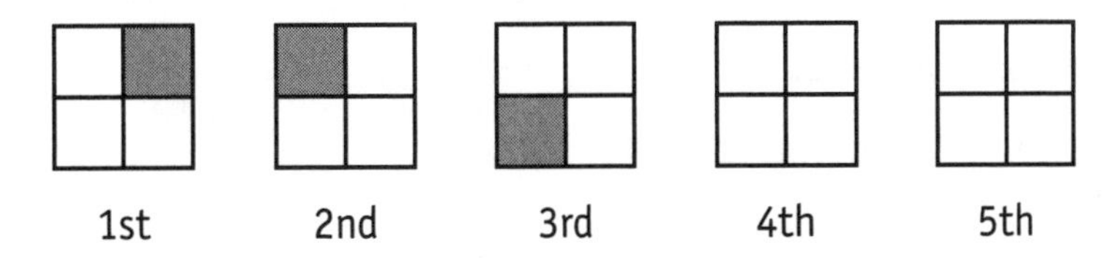

b. Shade the figure at the right to show the 100th figure in the shaded pattern.

100th

12. a. Draw the 4th and 5th figures of the pattern below.

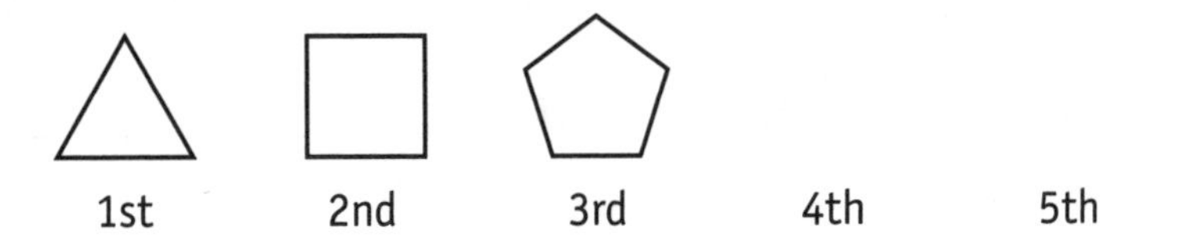

b. How many line segments are in the 100th figure in this pattern?

THE LANGUAGE OF ALGEBRA

Algebraic Expressions

Algebra is a tool that you can use to solve many types of problems. In algebra, letters are used to stand for numbers. The letters are called **variables** or **unknowns.** An **algebraic expression** contains both numbers and variables.

> **Did You Know . . . ?**
> Any letter can be used to stand for a number.

Algebraic expression: $x + 3$
↑ variable

In this expression, the letter x stands for a number. Because the value of x is not given, the value of the expression is not known.

An algebraic expression may contain one or more numbers and variables combined by one or more +, –, ×, and ÷ signs.

Algebraic Expression	Word Expression
$x + 5$	*x* plus 5
$y - 2$	*y* minus 2 *or y* subtract 2
$8 - z$	8 minus *z or* 8 subtract *z*
$4n$	4 times *n or* the product of 4 and *n*
$\frac{d}{2}$	*d* divided by 2 *or* one-half *d*

Write an algebraic expression for each word expression.

1. a number *n* plus 7 __________
2. *x* minus 14 __________
3. the product of 8 and *x* __________
4. *y* divided by 7 __________
5. 6 times *w* __________
6. *r* subtract 36 __________

Write a word expression for each algebraic expression.

7. $m - 16$ __________
8. $9z$ __________
9. $\frac{a}{4}$ __________
10. $4 + y$ __________
11. $p \div 10$ __________
12. $n + 8$ __________

Picturing an Algebraic Expression

An algebraic expression can often be represented by a picture.

EXAMPLE The distance $\overline{AC}$ is the sum of the distances $\overline{AB}$ and $\overline{BC}$. This sum can be written as an algebraic expression.

$\overline{AC} = \overline{AB} + \overline{BC} = \boldsymbol{x + 8}$

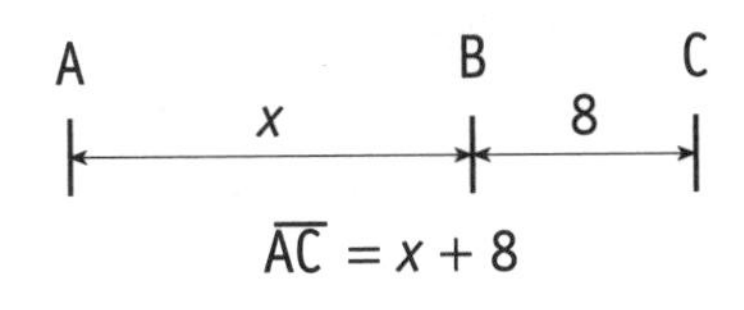

Becoming familiar with algebraic expressions is an important algebra skill.

Write an algebraic expression for each length. The first one in each row is done as an example.

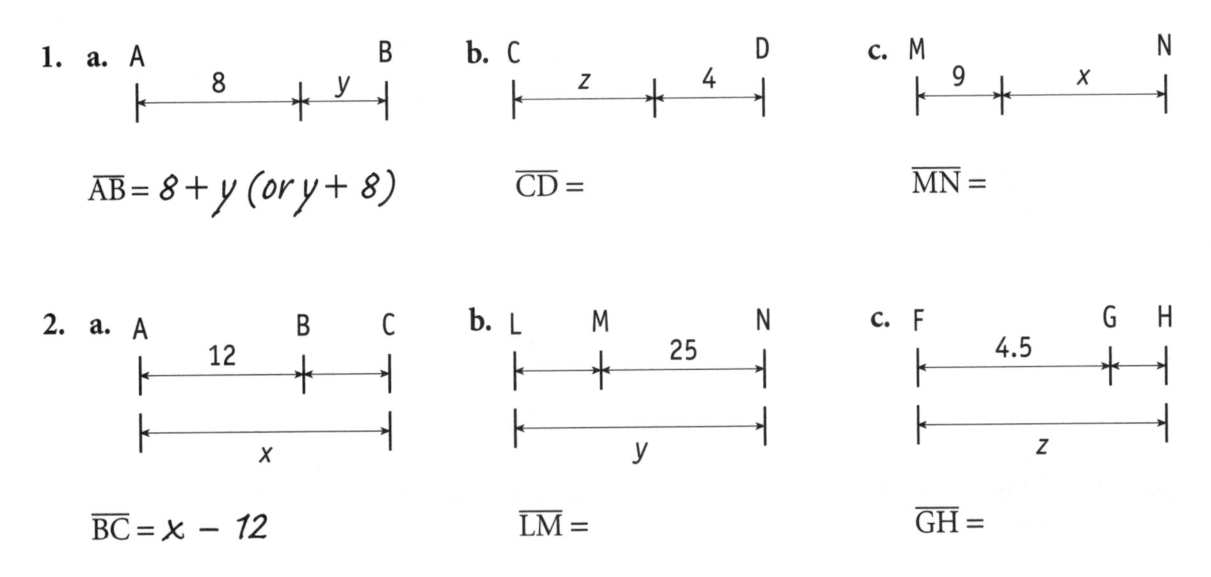

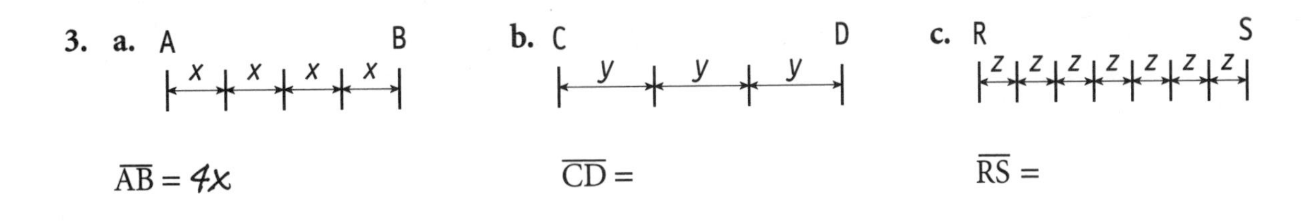

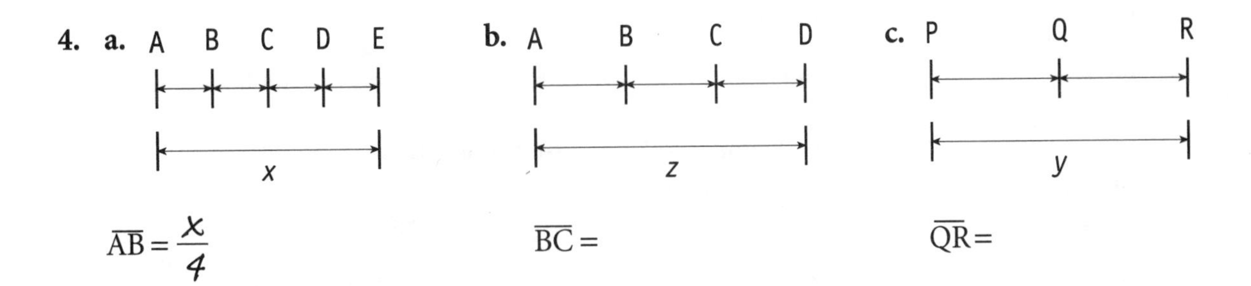

Word Problem Skills: Algebraic Expressions

An important skill in algebra is to be able to represent an amount as an algebraic expression.

EXAMPLE 1 Suppose m represents the money that Jill has in her savings account. Next month, Jill will place $25 more in this account. Write an algebraic expression that tells how much Jill will have saved after next month.

ANSWER: $m + \$25$

EXAMPLE 2 Four friends agree to share the cost of lunch. Let c stand for the total cost of lunch. Write an algebraic expression that tells each friend's share of the bill.

ANSWER: $\frac{c}{4}$

For problems 1–8, choose the algebraic expression that represents the word problem.

1. Ephran is now y years old. Which expression tells how old Ephran will be in 9 years?

   a. $y - 9$ **b.** $9 - y$ **c.** $y + 9$ **d.** $9y$

2. Before going to the store, Amy had m dollars. If she spent $3.75 at the store, which expression tells how much Amy has left?

   a. $m \times \$3.75$ **b.** $m - \$3.75$ **c.** $\$3.75 - m$ **d.** $\$3.75 + m$

3. Zander, Isaac, and Tim agree to share equally the money they earn on Saturday. If d stands for the total amount they earned, which expression tells each person's share?

   a. $3d$ **b.** $d + 3$ **c.** $d \div 3$ **d.** $d - 3$

4. The sales tax in Jenny's state is 5%. Which expression tells how much sales tax Jenny will pay for a coat that costs n dollars?

   a. $n - 5$ **b.** $n - 0.5$ **c.** $0.5n$ **d.** $0.05n$

5. Lucy and Demi bought sweaters. Demi paid $7 less than Lucy. If p stands for the price of Lucy's sweater, which expression tells the price of Demi's sweater?

   a. $p - \$7$ **b.** $p + \$7$ **c.** $\$7 - p$ **d.** $p \div 7$

6. Jackie's weight is two thirds the weight of her older sister Camille. If Camille weighs w pounds, which expression tells Jackie's weight?

   a. $w \div \frac{2}{3}$ b. $w + \frac{2}{3}$ c. $\frac{2}{3}w$ d. $\frac{2}{3} - w$

7. Leng bought a television on sale for 30% off. If the regular price of the set is r, which expression tells the amount Leng saved by buying at the sale price?

   a. $30r$ b. $0.3r$ c. $r - 30$ d. $r + 0.3$

8. The ratio of Caryn's age to Dusty's age is 3 to 2. If Dusty is d years old, which expression can be used to find Caryn's age?

   a. $\frac{3}{2}d$ b. $d \div \frac{3}{2}$ c. $\frac{3}{2} \div d$ d. $d + \frac{3}{2}$

For problems 9–14, write an algebraic expression as indicated.

9. Brendi earns $5.50 per hour. Suppose she gets a raise of r dollars per hour. Write an expression for her new hourly earnings.

10. A jacket, normally selling for n dollars, is on sale for 25% off. Write an expression that tells the amount of savings (s) being offered on the jacket.

11. Landon and three friends agree to share equally the cost (p) of a large pizza. Write an expression that tells the amount of each person's share.

12. Before he started his diet, Phil weighed w pounds. Phil now weighs 185 pounds. Write an expression that tells how much weight Phil has lost.

13. N students signed up for basketball. The students will be divided into six teams. Write an expression that tells about how many students will be on each team.

14. The ratio of men to women in Sally's art class is 1 to 4. There are 16 women in the class. Write a proportion that can be used to find the number (n) of men in the class.

Using Exponents

A **power** is the product of a number multiplied by itself one or more times. For example, four to the third power means "$4 \times 4 \times 4$." A power is usually written as a **base** and an **exponent.**

$4 \times 4 \times 4$ is written 4^3 ← exponent; ↑ base (the 4)

The exponent (^{3}) tells how many times to write the base (4) in the product.

- To find the value of a power, multiply and find the total.
- The value of 4^3 is $4 \times 4 \times 4 = 64$.

Product	As a Base and an Exponent	Word Expression	Value
5×5	5^2	five to the second power *or* five squared	25
$2 \times 2 \times 2$	2^3	two to the third power *or* two cubed	8
$10 \times 10 \times 10 \times 10$	10^4	ten to the fourth power	10,000
$\frac{1}{3} \times \frac{1}{3} \times \frac{1}{3}$	$(\frac{1}{3})^3$	one third cubed	$\frac{1}{27}$

(**Note:** A number raised to the second power is *squared.* A number raised to the third power is *cubed.* These are the only two powers that have special names.)

Write each product as a base and an exponent.

1. $2 \times 2 =$ $\qquad$ $3 \times 3 \times 3 \times 3 =$ $\qquad$ $\frac{1}{3} \times \frac{1}{3} =$

2. $6 \times 6 \times 6 =$ $\qquad$ $12 \times 12 =$ $\qquad$ $\frac{1}{2} \times \frac{1}{2} \times \frac{1}{2} \times \frac{1}{2} =$

3. $8 \times 8 \times 8 =$ $\qquad$ $9.2 \times 9.2 \times 9.2 =$ $\qquad$ $\frac{5}{8} \times \frac{5}{8} \times \frac{5}{8} \times \frac{5}{8} \times \frac{5}{8} =$

Find each value.

4. $7^2 =$ $\qquad$ $2^4 =$ $\qquad$ $(\frac{1}{3})^4 =$

5. $4^3 =$ $\qquad$ $1^3 =$ $\qquad$ $(\frac{3}{4})^3 =$

6. $5^2 =$ $\qquad$ $10^3 =$ $\qquad$ $(\frac{3}{8})^2 =$

Finding a Square Root

The opposite of squaring a number is finding a **square root.**

To find the square root of 36 you ask, "What number times itself equals 36?" The answer is 6 because $6 \times 6 = 36$.

The symbol for square root is $\sqrt{\ }$. The square root of 36 is 6 which can be written as $\sqrt{36} = 6$.

The table below contains the squares of numbers from 1 to 15. The numbers 1, 4, 9, and so on are called **perfect squares** because *their square roots are whole numbers.*

Table of Perfect Squares				
$1^2 = 1$	$4^2 = 16$	$7^2 = 49$	$10^2 = 100$	$13^2 = 169$
$2^2 = 4$	$5^2 = 25$	$8^2 = 64$	$11^2 = 121$	$14^2 = 196$
$3^2 = 9$	$6^2 = 36$	$9^2 = 81$	$12^2 = 144$	$15^2 = 225$

You can use the table to find the square roots of the first 15 perfect squares. For example, the table shows that the square root of 81 is 9 because $9^2 = 81$. Thus $\sqrt{81} = 9$.

You can also use this table to estimate a square root.

EXAMPLES $\sqrt{38} \approx 6$ because $38 \approx 36$ (**Remember:** $\approx$ means "is approximately equal to")

$\sqrt{30} \approx 5.5$ because 30 is about halfway between 25 and 36.

The square root of a fraction is equal to the square root of the numerator over the square root of the denominator.

EXAMPLES $\sqrt{\frac{4}{25}} = \frac{\sqrt{4}}{\sqrt{25}} = \frac{2}{5}$ $\qquad$ $\sqrt{\frac{9}{49}} = \frac{\sqrt{9}}{\sqrt{49}} = \frac{3}{7}$

Use the Table of Perfect Squares to find each square root.

1. $\sqrt{121} =$ $\qquad$ $\sqrt{64} =$ $\qquad$ $\sqrt{196} =$ $\qquad$ $\sqrt{49} =$

Find the square root of each fraction.

2. $\sqrt{\frac{4}{49}} =$ $\qquad$ $\sqrt{\frac{1}{9}} =$ $\qquad$ $\sqrt{\frac{25}{64}} =$ $\qquad$ $\sqrt{\frac{4}{16}} =$

Estimate each square root.

3. $\sqrt{21} \approx$ $\qquad$ $\sqrt{6.3} \approx$ $\qquad$ $\sqrt{55} \approx$ $\qquad$ $\sqrt{183} \approx$

Evaluating Algebraic Expressions

To **evaluate** (find the value of) an algebraic expression, replace each variable with a number, and then do the indicated operations.

EXAMPLE 1 Find the value of $x + 7$ when $x = 4$.

STEP 1 Substitute 4 for x. $x + 7 = 4 + 7$

STEP 2 Add 4 and 7. $= \mathbf{11}$

ANSWER: 11

EXAMPLE 2 Find the value of rt when $r = 65$ and $t = 3$.

STEP 1 Substitute 65 for r and 3 for t. (rt means $r \times t$) $rt = 65(3)$

STEP 2 Multiply 65 by 3. $= \mathbf{195}$

ANSWER: 195

When more than one operation is involved, multiply or divide *before* adding or subtracting.

EXAMPLE 3 Find the value of $3y - 6$ when $y = 8$.

STEP 1 Substitute 8 for y. $3y - 6 = 3(8) - 6$

STEP 2 Multiply 3 by 8. $= 24 - 6$

STEP 3 Subtract 6 from 24. $= \mathbf{18}$

ANSWER: 18

EXAMPLE 4 Find the value of $a^2 + 4a$ when $a = 3$.

STEP 1 Substitute 3 for a. $a^2 + 4a = 3^2 + 4(3)$

STEP 2 Evaluate 3^2 and $4(3)$. $= 9 + 12$

STEP 3 Add 9 and 12. $= 21$

ANSWER: 21

Evaluate each algebraic expression.

1. $x + 8$ when $x = 5$ $\qquad$ $y - 7$ when $y = 12$ $\qquad$ $z + \$3.50$ when $z = \$1.75$

2. rt when $r = 28, t = 1.5$ $\qquad$ $b^2 - b$ when $b = 3$ $\qquad$ $\frac{d}{r}$ when $d = 240, r = 60$

3. $3y + 4$ when $y = 7$ $\qquad$ $5x - 9$ when $x = 8$ $\qquad$ $\frac{a}{b} + 2ab$ when $a = 3, b = 4$

Find the perimeter of each figure.

4. $P =$ ______ $\qquad$ 5. $P =$ ______

$x = 6$ ft $\qquad$ $a = 6$ in., $b = 4$ in.

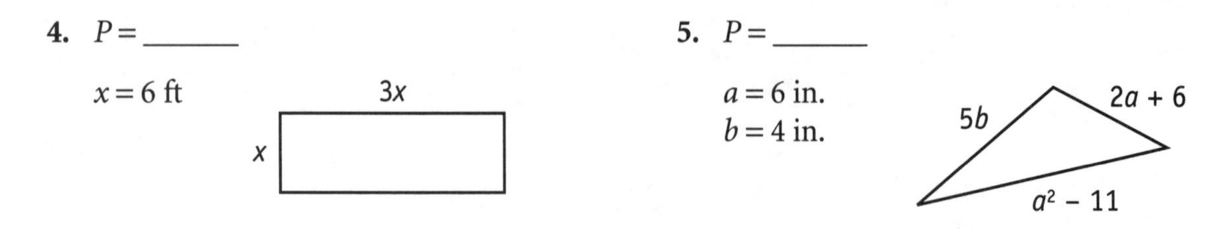

If an algebraic expression contains parentheses, find the value of the expression within the parentheses *before* performing any other operation.

EXAMPLE 5 Find the value of $6(x - 3)$ when $x = 5$.

STEP 1 Substitute 5 for x. $6(x - 3) = 6(5 - 3)$

STEP 2 Subtract 3 from 5. $= 6(2)$

STEP 3 Multiply 6 times 2. $= \mathbf{12}$

ANSWER: 12

EXAMPLE 6 Find the value of $2(s^2 + 5)$ when $s = 3$.

STEP 1 Substitute 3 for s. $2(s^2 + 5) = 2(3^2 + 5)$

STEP 2 Square 3. $= 2(9 + 5)$

STEP 3 Add 9 and 5. $= 2(14)$

STEP 4 Multiply 2 by 14. $= \mathbf{28}$

ANSWER: 28

Find the value of each algebraic expression.

6. $4(x - 3)$ when $x = 5$ $\quad$ $3(y + 2)$ when $y = 2$ $\quad$ $6(n - 1)$ when $n = 4$

7. $2(a^2 - 6)$ when $a = 3$ $\quad$ $4(r^2 - 1)$ when $r = 2$ $\quad$ $6(y^2 - 3)$ when $y = 4$

Find the value of the expression within parentheses *before* you find the square root.

8. $\sqrt{(x+21)}$ when $x = 4$ $\quad$ $\sqrt{(2y+2)}$ when $y = 7$ $\quad$ $\sqrt{(3a-14)}$ when $a = 26$

9. $\sqrt{(a^2+9)}$ when $a = 4$ $\quad$ $\sqrt{(a^2+b^2)}$ when $a = 6$, $b = 8$ $\quad$ $\sqrt{(c^2-b^2)}$ when $c = 13$, $b = 12$

Becoming Familiar with Equations

An **equation** is a statement that says two quantities are equal. You can write an equation in words or in symbols.

In words: seventeen minus nine equals eight
In symbols: $17 - 9 = 8$

Each time you add, subtract, multiply, or divide, you write an equation.

Familiar Equations	In Symbols	In Words
Addition equation:	$15 + 9 = 24$	fifteen plus nine equals twenty-four
Subtraction equation:	$26 - 18 = 8$	twenty-six minus eighteen equals eight
Multiplication equation:	$12 \times 5 = 60$	twelve times five equals sixty
Division equation:	$36 \div 9 = 4$	thirty-six divided by nine equals four

Basic Algebraic Equations

In an algebraic equation, a **variable** (letter) stands for an unknown number. The variable is usually referred to as the *unknown.* Here are examples of the four basic algebraic equations.

Algebraic Equations	In Symbols	In Words
Addition equation:	$x + 9 = 24$	x plus nine equals twenty-four
Subtraction equation:	$y - 18 = 8$	y minus eighteen equals eight
Multiplication equation:	$5n = 60$	five times n equals sixty
Division equation:	$\frac{x}{9} = 4$	x divided by nine equals four

Write each equation in symbols.

1. x minus forty-one equals thirty-three ____________
2. y plus seventeen equals twenty-nine ____________
3. n divided by eight equals nine ____________
4. six times z equals seventy-two ____________
5. x divided by fourteen equals seventy ____________

Picturing an Equation

The relationship expressed by an equation can often be represented by a picture.

EXAMPLE The distance between points A and B can be written in either of two ways:

$\overline{AB} = x + 6$ *or* $\overline{AB} = 19$

The expression $x + 6$ has the same value as the number 19. As an equation, this relationship is written $x + 6 = 19$

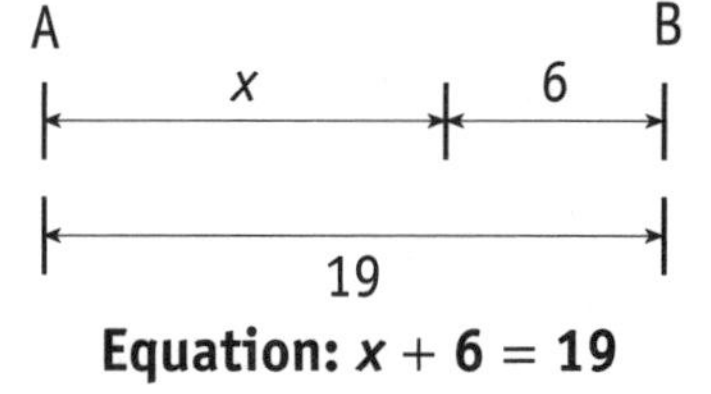

Equation: $x + 6 = 19$

Write an addition equation that expresses each relationship.

Equation:

Equation:

Write an subtraction equation that expresses each relationship.

Equation:

Equation:

Write a multiplication or division equation that expresses each relationship.

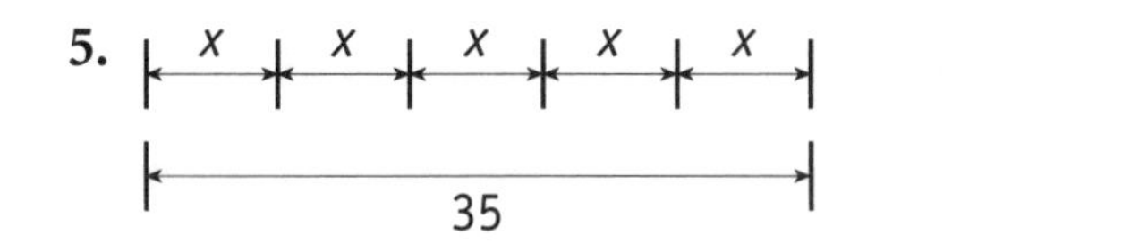

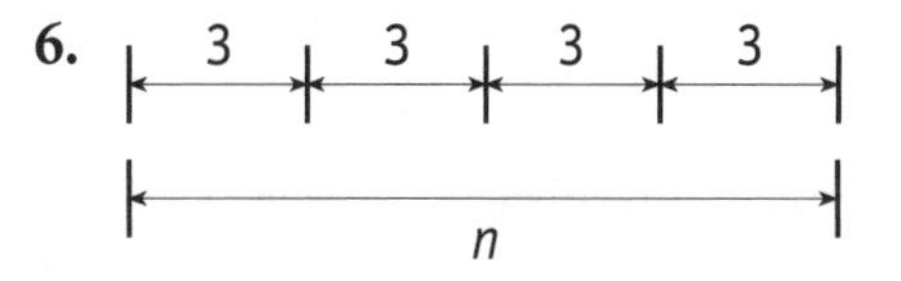

Equation:

Equation:

Solving an Equation

To solve an algebraic equation is to find the value of the unknown that makes the equation a true statement. Each equation you will study in this chapter has a single solution.

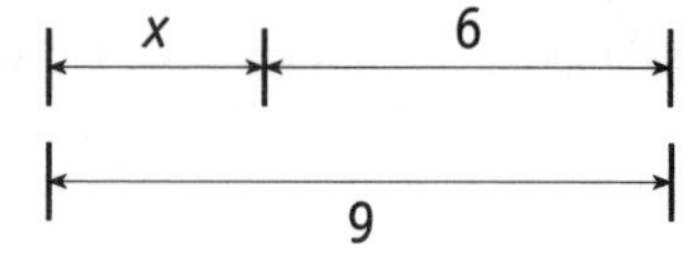

EXAMPLE For what value of x is the equation $x + 6 = 9$ true?

ANSWER: $\mathbf{x = 3}$ because $\mathbf{3} + 6 = 9$.

You can solve each type of basic algebraic equation by guessing the solution or you can use an **inverse operation.** An inverse operation "undoes" an operation and leaves the unknown standing alone.

Operation	Inverse Operation	Solving an Equation
addition	subtraction	Use subtraction to solve an addition equation.
subtraction	addition	Use addition to solve a subtraction equation.
multiplication	division	Use division to solve a multiplication equation.
division	multiplication	Use multiplication to solve a division equation.

STEP 1 Perform an inverse operation.

The equation $x + 6 = 9$ contains addition. To undo addition, subtract 6 from each side. Subtraction is the inverse (opposite) of addition.

$x + 6 - 6 = 9 - 6$

STEP 2 Simplify the equation.

To simplify the equation, complete each subtraction.

$x = 9 - 6$ (6 − 6 = 0, leaving x alone)

$\mathbf{x = 3}$ (9 − 6 = 3)

An equation is like a balanced scale.

An operation done on one side must also be done on the other side.

Guess the solution of each equation.

1. a. $x + 4 = 9$ **b.** $y - 2 = 6$ **c.** $2n = 14$ **d.** $\frac{x}{3} = 4$

Match each operation with its inverse.

_____ **2.** addition **a.** division

_____ **3.** subtraction **b.** subtraction

_____ **4.** multiplication **c.** addition

_____ **5.** division **d.** multiplication

Checking a Solution

Solving an equation means finding the value that makes the equation a true statement. The **solution** is the value of the unknown that solves the equation.

To check if a value of the unknown is the correct solution, follow these steps.

STEP 1 Substitute the value for the unknown into the equation.

STEP 2 Do the operations and compare each side of the equation.

EXAMPLE 1 Is $x = 6$ the solution of $2x - 5 = 7$?

STEP 1 Substitute 6 for *x*. $2(6) - 5 = 7$?

STEP 2 Subtract. Compare. $12 - 5 = 7$?

$7 = 7$?

Since 7 = 7, ***x* = 6 is the solution** of the equation.

EXAMPLE 2 Is $n = 4$ the solution for $n + 5 = 11$?

STEP 1 Substitute 4 for *n*. $4 + 5 = 11$?

STEP 2 Add. Compare. $9 = 11$?

Since 9 is not equal to 11, ***n* = 4 is not the solution** of the equation.

Check if the suggested value is the solution of the equation. Circle Yes if the value is correct, No if it is not.

1. $x + 4 = 7$ Try $x = 2$ Yes No

2. $n - 8 = 10$ Try $n = 18$ Yes No

3. $2y = 14$ Try $y = 8$ Yes No

4. $\frac{x}{5} = 3$ Try $x = 15$ Yes No

5. $\frac{2}{3}z = 8$ Try $z = 12$ Yes No

6. $x + (-3) = 13$ Try $x = 10$ Yes No

7. $x - (-5) = 12$ Try $x = 7$ Yes No

8. $3m = -15$ Try $m = -7$ Yes No

Solving an Addition Equation

To solve an addition equation, subtract the *added number* from each side of the equation. The unknown will be left alone, and the equation is solved.

EXAMPLE Solve for x: $x + 8 = 13$ **Solve:** $x + 8 = 13$

STEP 1 Subtract 8 from each side of the equation. $x + 8 - 8 = 13 - 8$

STEP 2 Simplify both sides of the equation.
On the left, $8 - 8 = 0$, leaving x alone.
On the right, $13 - 8 = 5$. $\mathbf{x = 5}$

ANSWER: $x = 5$

Check the answer by substituting 5 for x. **Check:** $5 + 8 = 13$
✓$13 = 13$

Solve each addition equation and check each answer. The first problem in each row is partially completed. Show all work.

1. $x + 3 = 9$ $\quad x + 5 = 8$ $\quad y + 7 = 15$ $\quad n + 11 = 25$

$x + 3 - 3 = 9 - 3$

2. $y + \$4 = \$11$ $\quad n + \$8 = \$19$ $\quad z + \$7 = \$23$ $\quad x + 27¢ = 45¢$

$y + \$4 - \$4 = \$11 - \$4$

3. $x + \frac{1}{2} = 3$ $\quad y + \frac{3}{4} = 4$ $\quad a + \frac{1}{3} = \frac{5}{3}$ $\quad n + 2\frac{3}{8} = 6\frac{7}{8}$

$x + \frac{1}{2} - \frac{1}{2} = 3 - \frac{1}{2}$

4. $p + 2.5 = 7.7$ $\quad x + 1.8 = 3.5$ $\quad y + 4.8 = 12$ $\quad a + 3.75 = 6.5$

$p + 2.5 - 2.5 = 7.7 - 2.5$

Solving a Subtraction Equation

To solve a subtraction equation, add the number being subtracted from each side of the equation. The unknown will be left alone, and the equation is solved.

EXAMPLE Solve for x: $x - 6 = 9$ **Solve:** $x - 6 = 9$

STEP 1 Add 6 to each side of the equation. $x - 6 + 6 = 9 + 6$

STEP 2 Simplify both sides of the equation.
On the left, $-6 + 6 = 0$, leaving x alone. $\mathbf{x = 15}$
On the right, $9 + 6 = 15$.

ANSWER: $\mathbf{x = 15}$

Check the answer by substituting 15 for x. **Check:** $15 - 6 = 9$
✓$9 = 9$

Solve each subtraction equation and check each answer. The first problem in each row is partially completed. Show all work.

1. $x - 4 = 11$ $x - 7 = 6$ $y - 9 = 11$ $p - 13 = 20$
$x - 4 + 4 = 11 + 4$

2. $y - \$3 = \$9$ $x - \$6 = \$12$ $n - \$12 = \$15$ $y - 15¢ = 32¢$
$y - \$3 + \$3 = \$9 + \$3$

3. $x - \frac{2}{3} = 4$ $m - \frac{3}{4} = 2$ $y - \frac{5}{8} = \frac{2}{8}$ $a - 1\frac{1}{2} = 2\frac{1}{2}$
$x - \frac{2}{3} + \frac{2}{3} = 4 + \frac{2}{3}$

4. $n - 1.5 = 3.6$ $x - 2.6 = 7$ $y - 5.3 = 4.7$ $r - 2.25 = 7.9$
$n - 1.5 + 1.5 = 3.6 + 1.5$

Word Problem Skills [+ and −]

On this page, you'll see how simple algebraic equations can be written for problems you can easily solve without algebra. Practicing algebra skills here, though, is an excellent way to become more confident with the use of equations.

To solve a word problem using algebra, follow these steps.

STEP 1 Assign a letter to represent the unknown amount.

STEP 2 Write an equation that represents the given information.

STEP 3 Solve the equation to find the unknown value.

EXAMPLE Jason sold his car for $7,200. He received $2,650 less than he paid for the car 2 years ago. How much did Jason pay for the car?

STEP 1 Let x represent the amount Jason paid.

STEP 2 Write an equation. $x - \$7,200 = \$2,650$

STEP 3 Solve the equation. $x - \$7,200 = \$2,650$

Add $7,200 to each side. $x - \$7,200 + \$7,200 = \$2,650 + \$7,200$

ANSWER: $9,850 $\mathbf{x = \$9,850}$

Choose the correct equation, and then find the value of the unknown.

1. If you add 41 to a number, the sum is 53. If n stands for the unknown number, which equation can be used to find n?

   a. $n + 53 = 41$
   b. $n + 41 = 53$
   c. $n - 41 = 53$
   $n =$ ________

2. Evan paid $25 for a sweater that usually costs $40. Which equation can be used to find the amount Evan saved (s)?

   a. $s - \$25 = \$40$
   b. $s - \$40 = \$25$
   c. $s + \$25 = \$40$
   $s =$ ________

3. The temperature in Phoenix has risen 25°F to 112°F since noon. Which equation can be used to find the temperature at noon (t)?

   a. $t + 112°F = 25°F$
   b. $t - 112°F = 25°F$
   c. $t + 25°F = 112°F$
   $t =$ ________

4. Esther spent $4.50 for lunch. She now has $1.75. Which equation can be used to find the amount Esther had before lunch (m)?

   a. $m - \$4.50 = \$1.75$
   b. $m - \$1.75 = \$4.50$
   c. $m = \$4.50 - \$1.75$
   $m =$ ________

Solving a Multiplication Equation

To solve a multiplication equation, divide each side of the equation by the number that multiplies the unknown. The unknown will be left, and the equation is solved.

EXAMPLE Solve for x: $6x = 42$

Solve: $6x = 42$

STEP 1 Divide the value on each side of the equation by 6.

$$\frac{{}^{1}\not{6}x}{\not{6}_{1}} = \frac{42}{6}$$

STEP 2 Simplify both sides of the equation. On the left, $\frac{6}{6} = 1$. On the right, $\frac{42}{6} = 7$.

$$x = 7$$

ANSWER: $x = 7$

Check the answer by substituting 7 for x.

Check: $6(7) = 42$

✓$42 = 42$

Solve each multiplication equation and check each answer. The first problem in each row is partially completed. Show all work.

1. $5x = 25$ $\quad \frac{5x}{5} = \frac{25}{5}$ $\qquad 8n = 48 \qquad 6y = 54 \qquad 7m = 49$

2. $3z = 42$ $\quad \frac{3z}{3} = \frac{42}{3}$ $\qquad 5x = 65 \qquad 6a = 72 \qquad 9y = 126$

3. $4y = \$12$ $\quad \frac{4y}{4} = \frac{\$12}{4}$ $\qquad 7x = \$63 \qquad 8p = \$120 \qquad 12y = 96¢$

4. $3n = 15.6$ $\quad \frac{3n}{3} = \frac{15.6}{3}$ $\qquad 4y = 22.4 \qquad 7x = 44.8 \qquad 14z = 154$

Solving a Division Equation

To solve a division equation, multiply each side of the equation by the number that divides the unknown. The unknown will be left alone, and the equation will be solved.

EXAMPLE Solve for x: $\frac{x}{5} = 35$ **Solve:** $\frac{x}{5} = 35$

STEP 1 Multiply the value on each side of the equation by 5. $\frac{x}{5}(5) = 35(5)$

STEP 2 Simplify both sides of the equation. On the left, $\frac{5}{5} = 1$. On the right, $35(5) = 175$. $\mathbf{x = 175}$

ANSWER: $\mathbf{x = 175}$

Check the answer by substituting 175 for x. **Check:** $\frac{175}{5} = 35$

✓$35 = 35$

Solve each division equation and check each answer. The first problem in each row is partially completed. Show all work.

1. $\frac{x}{7} = 3$ $\qquad \frac{y}{4} = 20$ $\qquad \frac{z}{6} = 12$ $\qquad \frac{s}{9} = 8$

   $\frac{x}{7}(7) = 3(7)$

2. $\frac{y}{3} = \$5$ $\qquad \frac{m}{2} = \$17$ $\qquad \frac{x}{8} = \$11$ $\qquad \frac{r}{12} = \$2.50$

   $\frac{y}{3}(3) = \$5(3)$

3. $\frac{x}{2} = 2\frac{1}{2}$ $\qquad \frac{y}{3} = 2\frac{2}{3}$ $\qquad \frac{p}{5} = 3.5$ $\qquad \frac{n}{4} = 1.75$

   $\frac{x}{2}(2) = 2\frac{1}{2}(2)$

Word Problem Skills [× and ÷]

On this page, you'll practice using algebraic equations to solve multiplication and division problems.

EXAMPLE Kaitlyn lives in a state that has a 5% sales tax. On Saturday she bought a sweater on sale. She paid a sales tax of \$2.40. What was the sale price of the sweater?

STEP 1 Let s equal the sale price of the sweater.

STEP 2 Write an equation for s. $0.05s = \$2.40$

STEP 3 Solve the equation. $0.05s = \$2.40$

Divide each side by 0.05. $\frac{0.05}{0.05}s = \frac{\$2.40}{0.05}$

ANSWER: \$48.00 $s = \$48.00$

Choose the correct equation, and then find the value of the unknown.

1. If you multiply a number (n) by 12, the product is 204. Which equation below can be used to find n?

   a. $204n = 12$
   b. $12n = 204$
   c. $\frac{n}{12} = 204$

   $n =$ ______

2. One out of every 6 students in Quan's art class also takes Spanish. If 12 students in the art class take Spanish, which equation can be used to find the number of students (n) in the class?

   a. $\frac{n}{6} = 12$
   b. $6n = 12$
   c. $\frac{n}{12} = 6$

   $n =$ ______

3. Chris saves one-fifth of his monthly income. If he saves \$350 per month, which equation can be used to find his monthly income (i)?

   a. $5i = \$350$
   b. $i = \$350 \div 5$
   c. $\frac{1}{5}i = \$350$

   $i =$ ______

4. Denny paid a sales tax of \$3.60 when he bought a new jacket. If Denny paid a 6% sales tax, which equation can be used to find the cost (c) of the jacket?

   a. $\frac{c}{0.06} = \$3.60$
   b. $0.06c = \$3.60$
   c. $\$3.60c = 0.06$

   $c =$ ______

Reviewing One-Step Equations

You've now learned to solve four types of algebraic equations. Review your skills by solving these problems. Show all steps of each solution.

1. $x + 12 = 19$ $\quad y - 8 = 6$ $\quad \frac{n}{6} = 4$ $\quad 7z = 140$

2. $\frac{y}{3} = 5$ $\quad 3m = 18$ $\quad y + 3 = 12$ $\quad x - 12 = 20$

3. $4n = 32$ $\quad y + 21 = 21$ $\quad x - 11 = 19$ $\quad \frac{d}{2} = 14$

4. $x + \frac{2}{3} = 4$ $\quad y - \frac{1}{4} = 2$ $\quad \frac{p}{6} = \frac{2}{3}$ $\quad 4n = \frac{1}{2}$

5. $y + \frac{3}{4} = 2\frac{1}{4}$ $\quad n - \frac{2}{3} = 4\frac{1}{3}$ $\quad \frac{p}{2} = 4\frac{1}{2}$ $\quad 2x = 2\frac{1}{3}$

6. $x + 2.4 = 8.3$ $\quad y - 6.25 = 9$ $\quad \frac{r}{3} = 2.07$ $\quad 4p = 12.08$

For each problem below, do the following.

STEP 1 Choose a letter (*x, y, n, p,* and so on) to represent the unknown.

STEP 2 Write an equation to represent the information given in the problem.

STEP 3 Solve the equation.

The first problem is completed as an example. In each problem, there is more than one way to write a correct equation. Each correct equation has the same solution.

7. Wynton bought a jacket on sale for $65, $28 below its original price. What was the original price of the jacket?

a. Unknown: p = original price

b. Equation: $p - \$28 = \$65$

c. Solution: $p - \$28 + \$28 = \$65 + \$28$

$p = \$93$

$$\begin{array}{r} \overset{1}{\$65} \\ +\ 28 \\ \hline \$93 \end{array}$$

8. After growing 2.5 inches this year, Myra is now 65.5 inches tall. How tall was Myra at the beginning of this year?

a. Unknown:

b. Equation:

c. Solution:

9. Willema earns money by baby-sitting. She spends $\frac{1}{4}$ of her earnings on clothes. If she spent $240 this year on clothes, how much money did Willema earn during the year?

a. Unknown:

b. Equation:

c. Solution:

10. Janessa cut a ribbon into six equal pieces. If each piece is $1\frac{1}{2}$ feet long, how long was the ribbon before Janessa cut it?

a. Unknown:

b. Equation:

c. Solution:

11. After running 7.25 kilometers of a race, Manuel still has 2.75 kilometers to go. What distance is the race?

a. Unknown:

b. Equation:

c. Solution:

12. One third of the students in art class are boys. If eight students in the class are boys, how many students are in the class?

a. Unknown:

b. Equation:

c. Solution:

The Language of Algebra Review

Work each problem and check your answers. Correct any errors.

Write an algebraic expression for each word expression.

1. 7 times w

2. x minus 13

3. 16 divided by s

Write a word expression for each algebraic expression.

4. $y + 9$

5. $12 - x$

6. $7s$

For problems 7 and 8, choose the correct algebraic expression.

7. Nora, Leanne, and Wanda share equally money earned house cleaning on Saturday. If together they earned m dollars, which expression tells each person's share?

a. $m + 3$ **b.** $3m$ **c.** $\frac{m}{3}$ **d.** $\frac{3}{m}$

8. The ratio of girls to boys in Shari's dance class is 4 to 3. If there are n boys in the class, which expression can be used to find the number of girls?

a. $\frac{4}{3}n$ **b.** $\frac{3}{4}n$ **c.** $\frac{4}{3} + n$ **d.** $n - \frac{4}{3}$

For problems 9 and 10, write an algebraic expression as indicated.

9. Last year, Paul earned n dollars per hour baby-sitting. This year, he will earn $1.25 more each hour. Write an expression that tells how much money Paul will earn each hour he baby-sits this year.

10. The price of a pair of running shoes is on sale for 35% off the normal price (p). Write an expression that tells how much money can be saved by buying the shoes at the sale price.

Evaluate each algebraic expression.

11. $3x - 5$ when $x = 6$ $\quad$ rt when $r = 40, t = 1.5$ $\quad$ $3(x - 1)$ when $x = 4$

12. $3(a^2 - 11)$ when $a = 4$ $\quad$ $\sqrt{(x^2 + 36)}$ when $x = 8$ $\quad$ $\sqrt{(c^2 - b^2)}$ when $c = 10, b = 8$

Write an equation that expresses each relationship.

Equation: $\quad$ Equation:

Solve each equation. Show all steps.

15. $b + 9 = 16$ $\quad$ **16.** $x - 7 = 8$ $\quad$ **17.** $4n = 24$ $\quad$ **18.** $\frac{r}{3} = 12$

For problems 19 and 20, do the following.

- **Write an equation to represent the given information.**
- **Solve the equation, showing all steps.**

19. Jason saves 20% of his earnings. If he saved \$45 last month, how much money (m) did Jason earn?

Equation:

Solution:

20. One-third of Melanie's swimming class are girls. If seven girls are in the class, how many students (n) are in the class?

Equation:

Solution:

Pre-Algebra Posttest A

This posttest gives you a chance to check your pre-algebra skills. Take your time and work each problem carefully. When you finish, check your answers and review any topics on which you need more work.

1. On a number line, how many units to the left of 5 is –2?

2. Circle each possible value of n for the inequality $n \geq -3$.

   –5 –2 0 –9 4

3. Graph the following inequality on the number line.

   $y \geq -2$

   –6 –5 –4 –3 –2 –1 0 1 2 3 4 5 6

4. Find the value of the expression $5(7 - 3)$.

5. At a weekend sale, Dana sees a pair of shoes reduced from $48.95 to $34.49. *Estimate* how much Dana can save by buying the shoes at this sale.

6. Brandon is buying a television set on sale for $250. The sales tax in Brandon's state is 6%. Which expression tells how much Brandon will pay in all?

   a. $0.06 \times \$250$
   b. $\$250 + 0.06 \times \$250$
   c. $\$250 - 0.06 \times \$250$
   d. $\$250 + \$250 \div 0.06$

7. At Jefferson Elementary School, there are 20 teachers. If 14 of the teachers are women, what is the ratio of women teachers to men teachers?

8. In a survey, 2 out of 3 of 450 students said they watch a movie at least once each month. Write a proportion that can be used to find the number (n) of students who do *not* watch a movie at least once each month.

9. Janice is decorating a large picnic table. She will place a flower at each corner of the table. She will also place a flower every 2 feet on each side of the table. Along the center line of the table she will place a flower every foot. If the table is 4 feet wide and 10 feet long, how many flowers will Janice need?

A Drawing May Help

center line

10. Jeff, Roger, and Dustin are planning to go to a movie. How many different seating arrangements are possible? Assume the three friends sit next to each other in seats A, B, and C.

A List May Help

Seat A **Seat B** **Seat C**

11. Amanda sold eight posters and collected $37.25. She sold large posters for $5.50 and small posters for $3.25. How many small posters did Amanda sell?

A Table May Help

Small Poster $3.25	Large Poster $5.50	Total Cost
1	7	
2	6	
3		

Guess and Check May Help

12. Bruce, Colin, and Stewart are going camping. Their three backpacks together weigh 73 pounds. Colin's backpack is 1 pound heavier than Bruce's, Bruce's is 3 pounds heavier than Stewart's. What is the weight of each backpack?

Colin's: ____ pounds

Bruce's: ____ pounds

Stewart's: ____ pounds

For problems 13 and 14, refer to the line graph at the right.

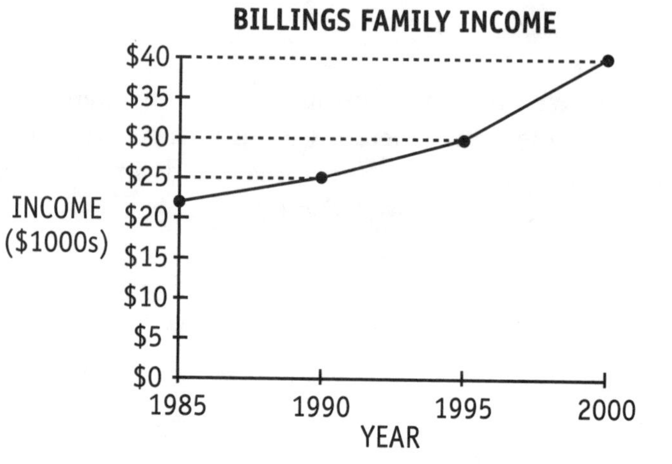

13. By about how much (in dollars) did the Billings family income increase between 1985 and 2000?

14. *Estimate* the Billings family income for the year 1997.

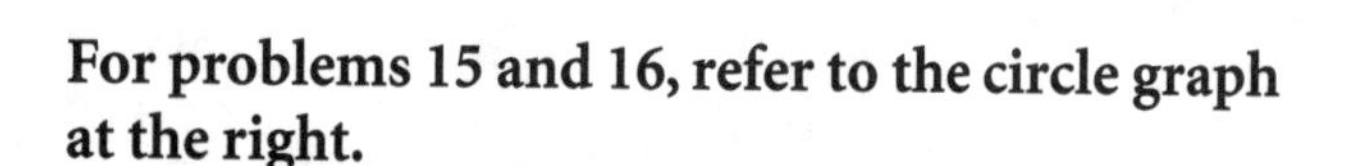

For problems 15 and 16, refer to the circle graph at the right.

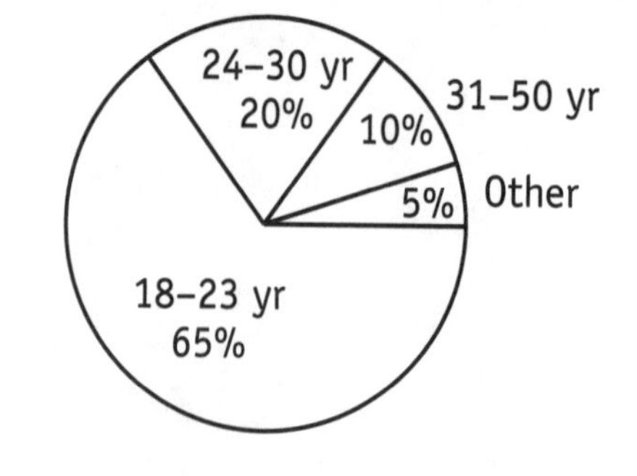

15. What is the ratio of first-year students in the 31–50 year age group to those in the 24–30 year age group?

16. How many of Lynn's 3,500 first-year students are in the 18–23 year age group?

17. The number cube at the right has six faces. With one toss, what is the probability of rolling a number that is greater than 2?

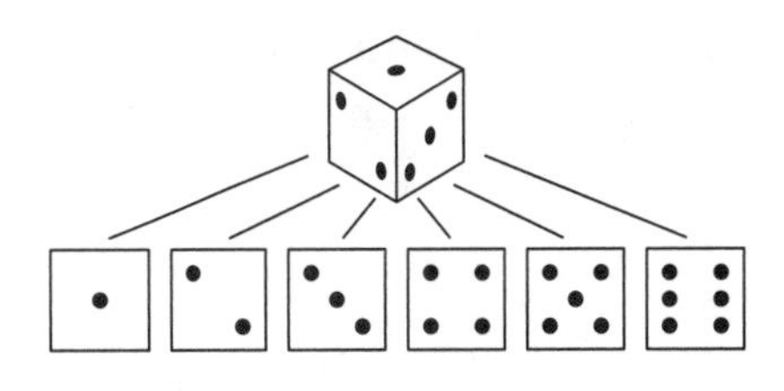

For problems 18 and 19, refer to the five cards shown below.

18. If you randomly choose one card, what is the probability that the card you choose will be a face card?

19. Suppose you choose a card, look at it, and return it. If you do this 50 times, how many times are you most likely to choose a number card?

20. Each ring of the dartboard has the same area. Each time Caren throws a dart, she is equally likely to hit any ring. If Caren throws two darts, what is the probability that both darts will land in the shaded ring?

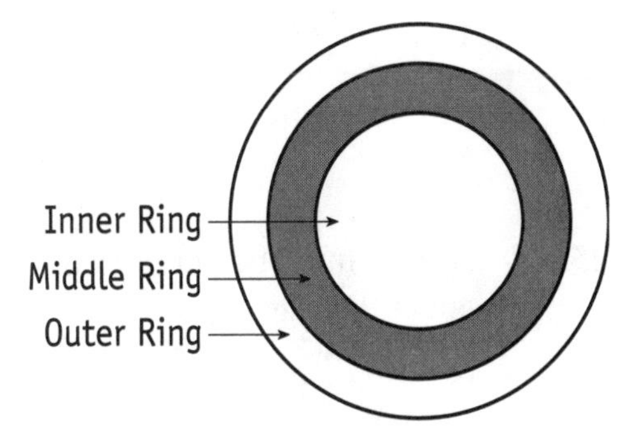

21. Angles C and D are supplementary angles. If the measure of $\angle C$ is 85°, what is the measure of $\angle D$?

22. In ΔABC, the measures of $\angle B$ and $\angle C$ are given. What is the measure of $\angle A$?

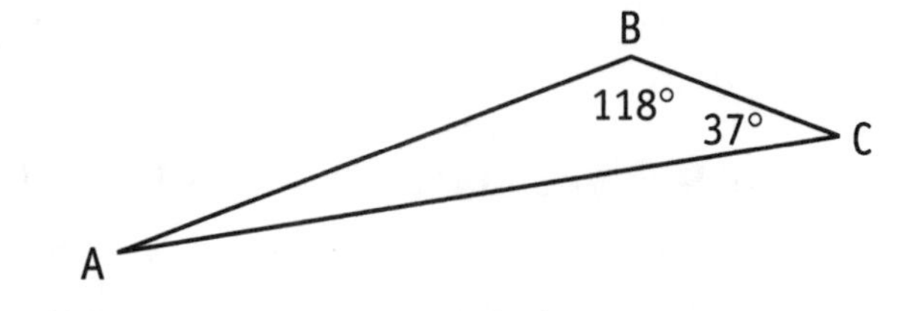

23. A dog is tied to a stake in the center of a yard. If the rope is 16 feet long, what is the area over which the dog can play? Express your answer to the nearest square foot.

16 ft

24. An aquarium is 1.5 feet wide, 3.25 feet long, and 1.625 feet high.

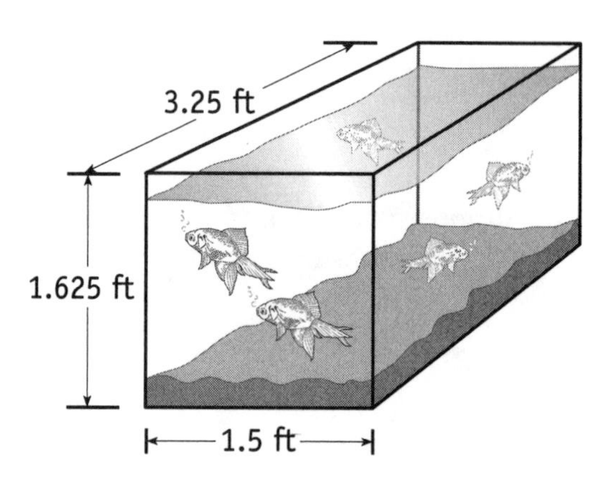

a. To the nearest 0.1 cubic foot, what is the volume of the aquarium?

b. To the nearest gallon, how much water is needed to fill the aquarium? (1 cubic foot ≈ 7.5 gallons)

25. Draw each line of symmetry for the equilateral triangle shown at the right.

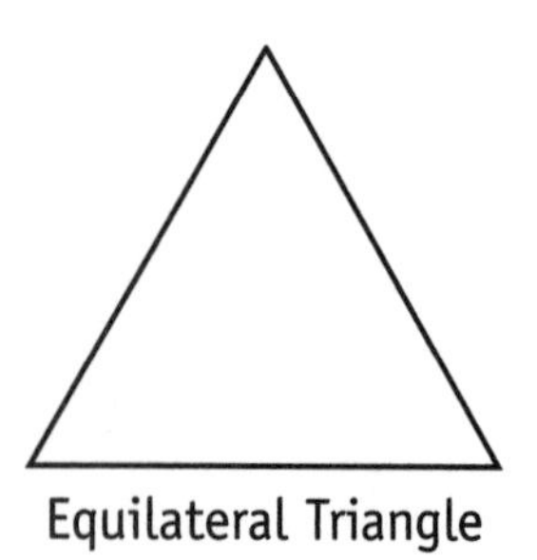

26. **a.** Draw two polygons that are congruent.

b. Draw two polygons that are similar but not congruent.

For problem 27, refer to the grid at the right.

27. Point D (not shown) is the fourth corner of a rectangle. The other three corners are points A, B, and C. What are the coordinates of point D?

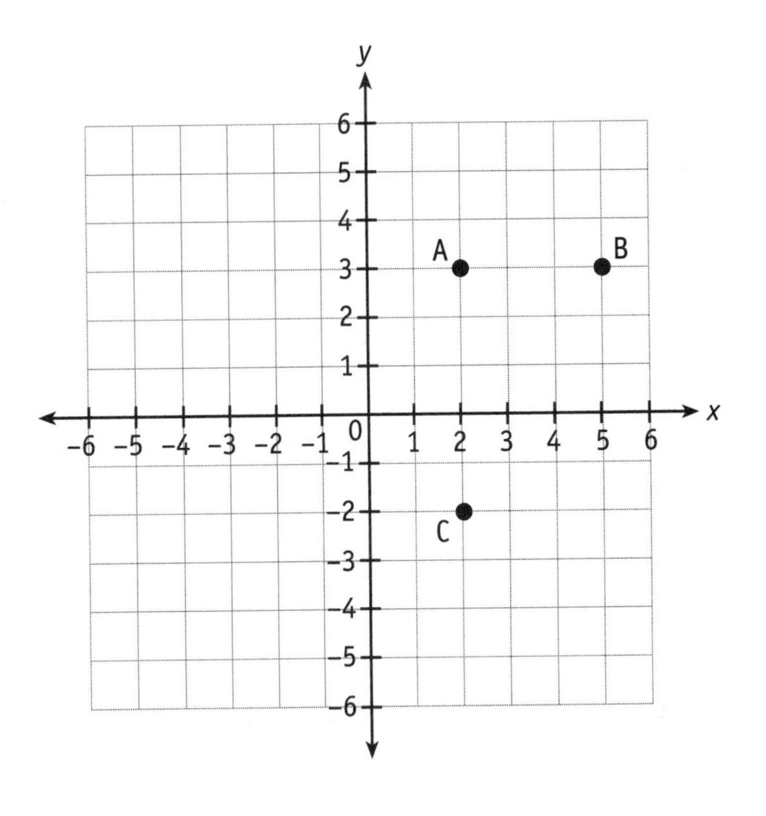

28. How many 60° angles are in the 4th figure in the pattern below? (**Hint:** As a first step, draw the 4th figure.)

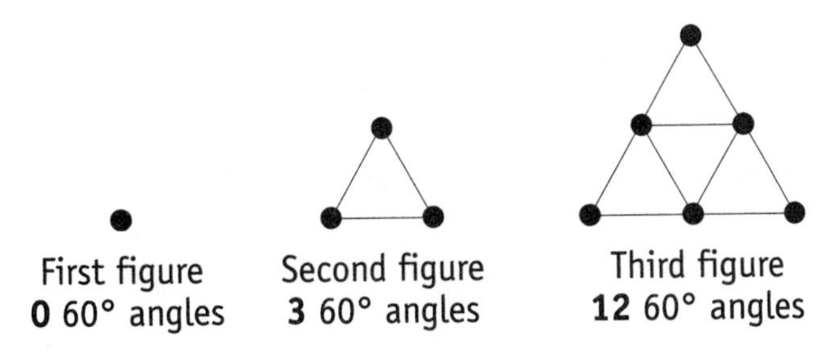

29. An umbrella that normally sells for $14.00 is on sale for $25% off. Write an algebraic expression that represents the sale price.

30. What is the value of the algebraic expression below when $x = 2$, $y = 3$, and $z = -1$?

$5x - y^2 + 4z^2$

31. Solve the addition equation below. Show all steps.

$x + 9 = 22$

32. Solve the division equation below. Show all steps.

$\frac{y}{4} = 9$

Pre-Algebra Posttest A Prescriptions

Circle the number of any problem that you miss. A passing score is 27 correct answers. If you passed the test, go on to Using Number Power. If you did not pass the test, review the chapters in this book or refer to these practice pages in other materials from Contemporary Books.

PROBLEM NUMBERS	SKILL AREA	PRACTICE PAGES
1, 2, 3, 4	number skills	10–29
	Math Exercises: Decimals/Percents/Fractions	3–27 in each book
	Real Numbers: Estimation 1 and 2	1–68
	Breakthroughs in Math: Book 2	34–140
5, 6, 7, 8	word problem skills	32–51
	Math Exercises: Problem Solving & Applications	3–26
9, 10, 11, 12	problem-solving strategies	54–67
	Critical Thinking with Math	26–51
13, 14, 15, 16	data analysis	70–85
	Math Exercises: Data Analysis & Probability	3–19
	Real Numbers: Tables, Graphs, & Data Interpretation	1–65
17, 18, 19, 20	probability	88–103
	Math Exercises: Data Analysis & Probability	20–26
21, 22, 23, 24	geometry	106–121
	Math Exercises: Geometry	3–27
	Real Numbers: Geometry Basics	1–68
25, 26, 27, 28	spatial sense and patterns	124–137
29, 30, 31, 32	algebra	140–159
	Math Exercises: Pre-Algebra	3–27
	Real Numbers: Algebra Basics	1–67

For further pre-algebra practice:
Math Solutions (software)
all units

Pre-Algebra Posttest B

This posttest gives you a chance to check your pre-algebra skills in a multiple-choice format as used in many standardized tests. Take your time and work each problem carefully before you choose your answer.

1. On a number line, how many units to the right of –5 is +5?

   a. 0 b. 2 c. 5 d. 6 e. 10

2. Which number is *not* a possible value of n in the inequality $-3 < n \leq 1$?

   a. 1 b. 0 c. –1 d. –2 e. –3

3. Which inequality is graphed on the number line?

   Values of x

   –6 –5 –4 –3 –2 –1 0 1 2 3 4 5 6

   a. $x \leq -3$ b. $x \geq -3$ c. $x < -3$ d. $x > -3$ e. $x = -3$

4. What is the value of the expression $4(5 + 3) - 4$?

   a. 32 b. 28 c. 19 d. 14 e. 8

5. What is the best estimate of the product $\$3.19 \times 19$?

   a. $\$3.00 \times 20$ b. $\$3.25 \times 20$ c. $\$3.50 \times 20$ d. $\$3.00 \times 25$ e. $\$3.25 \times 25$

6. Keone made a \$25 down payment on a \$170 bicycle. He will pay off the balance in six equal payments. Not counting any finance charge, which expression tells the amount of each payment?

   a. $(\$170 - \$25) \div 6$
   b. $(\$170 - \$25) \div 5$
   c. $(\$195 - \$25) \div 6$
   d. $(\$170 + \$25) \div 5$
   e. $(\$195 + \$25) \div 6$

7. In January Shauna's basketball team won 5 of the 8 games they played. What is the ratio of the games Shauna's team won to the games they lost?

   a. $\frac{3}{8}$ b. $\frac{3}{5}$ c. $\frac{8}{5}$ d. $\frac{5}{3}$ e. $\frac{5}{8}$

8. In a survey of 65 shoppers, 3 out of every 5 said they regularly ride the city bus. Which proportion can be used to find the number (n) of these shoppers who do *not* regularly ride the city bus?

a. $\frac{n}{65} = \frac{3}{5}$

b. $\frac{n}{65} = \frac{2}{5}$

c. $\frac{n}{65} = \frac{3}{7}$

d. $\frac{n}{65} = \frac{3}{8}$

e. $\frac{n}{65} = \frac{5}{2}$

9. Shelly, Kira, Dustin, and Aaron are going to play in a chess tournament. Each student will play each other one game. How many games will be played in all?

A List May Help

a. 6
b. 8
c. 10
d. 12
e. 16

10. Blake paid $4.75 for eight candy bars. Some of the bars cost 75¢ and some cost 50¢. How many of the more expensive candy bars did Blake buy?

a. 2
b. 3
c. 5
d. 6
e. 7

A Table May Help

50¢ Bars	75¢ Bars	Total Cost

11. Leanne's teacher placed a full box of colored pencils on the table. Leanne took 12 pencils and left the rest. Shari took half the pencils that Leanne left. Mark took half of what Shari didn't take, leaving 3 pencils in the box. How many pencils were in the full box?

a. 16
b. 18
c. 20
d. 22
e. 24

12. Tickets for games at Family Fun Night are on sale at Jefferson School. As shown in the table, the price per ticket decreases if you buy several at one time. If the price pattern continues, what is the total price of a packet of 10 tickets?

a. $6.00
b. $6.50
c. $7.00
d. $7.50
e. $8.00

Number of Tickets	Price per Ticket
2	90¢
4	85¢
6	80¢
8	
10	

For problems 13 and 14, refer to the pictograph at the right.

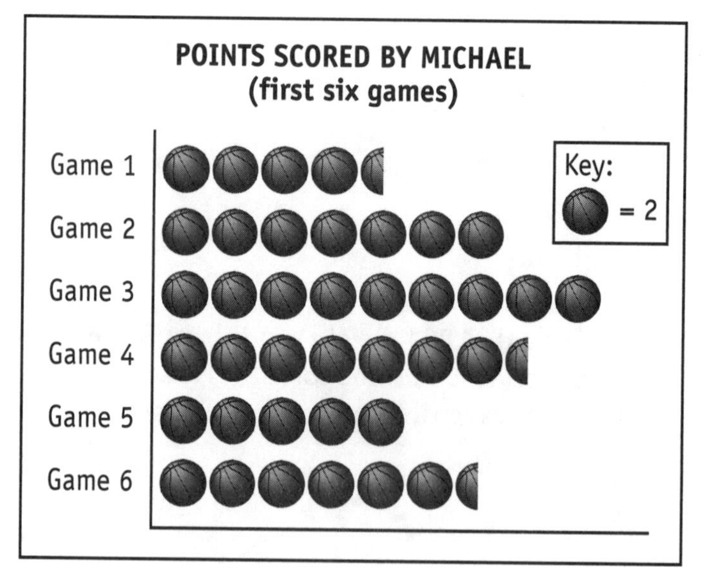

13. What is the difference between the most points Michael scored in a game and the least points he scored?

a. 7
b. 9
c. 12
d. 14
e. 18

14. To the nearest point, how many points did Michael average per game for these first six games?

a. 19
b. 16
c. 15
d. 13
e. 10

For problems 15 and 16, refer to the bar graph at the right.

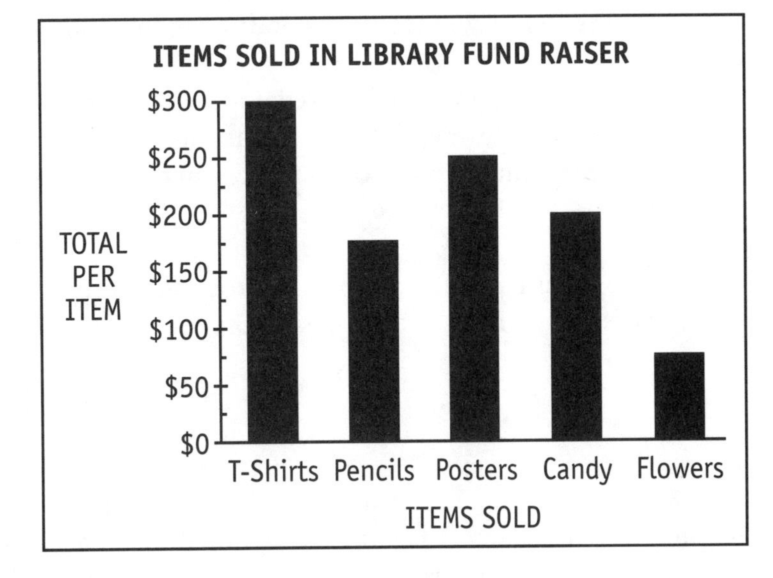

15. What is the ratio of money from sales of flowers to sales of pencils?

a. 1 to 2
b. 2 to 5
c. 3 to 7
d. 4 to 9
e. 5 to 11

16. What percent of the total sales came from the sales of posters?

a. 25%
b. 35%
c. 40%
d. 50%
e. 75%

For problems 17 and 18, refer to the spinner below, which is divided into five equal parts. Assume that the pointer does not land on a line.

17. What is the probability that, on one spin, the pointer will land on a shaded section?

a. $\frac{1}{2}$
b. $\frac{3}{7}$
c. $\frac{3}{5}$
d. $\frac{2}{7}$
e. $\frac{2}{5}$

18. Suppose you spin the pointer 50 times. Which phrase *best* describes the number of times the pointer is likely to land on a shaded section?

a. more than 30
b. fewer than 10
c. fewer than 25
d. exactly 25
e. more than 25

For problems 19 and 20, refer to the following information.

Stacey plays on the high school basketball team. During the first ten games of the season, Stacey made 15 free throws out of 20 tries. She also made 20 three-point shots out of 50 tries. Tonight, the team is playing its eleventh game.

19. Expressed as a percent, what is the probability that Stacey will make the first free throw she attempts in tonight's game?

a. 25% **b.** $33\frac{1}{3}\%$ **c.** 50% **d.** $62\frac{1}{2}\%$ **e.** 75%

20. Expressed as a fraction, what is the probability that Stacey will make the *first two* three-point shots she tries in tonight's game?

a. $\frac{3}{5}$ **b.** $\frac{2}{5}$ **c.** $\frac{3}{10}$ **d.** $\frac{4}{25}$ **e.** $\frac{1}{10}$

21. Angles A and B are complementary angles. If the measure of ∠A is 35°, what is the measure of ∠B?

a. 55° **b.** 90° **c.** 145° **d.** 235° **e.** 325°

22. In ΔCDE, the measures of ∠C and ∠D are given. What is the measure of ∠E?

a. 10°
b. 23°
c. 42°
d. 73°
e. 90°

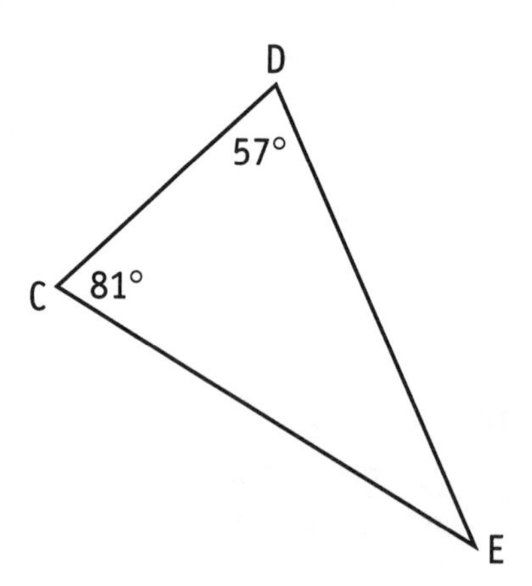

23. A sprinkler sends out a spray of water that reaches a point 12 feet from the sprinkler head. If the sprinkler slowly turns in a circle, what is the approximate area of lawn that is watered?

a. 40 sq ft
b. 120 sq ft
c. 250 sq ft
d. 450 sq ft
e. 780 sq ft

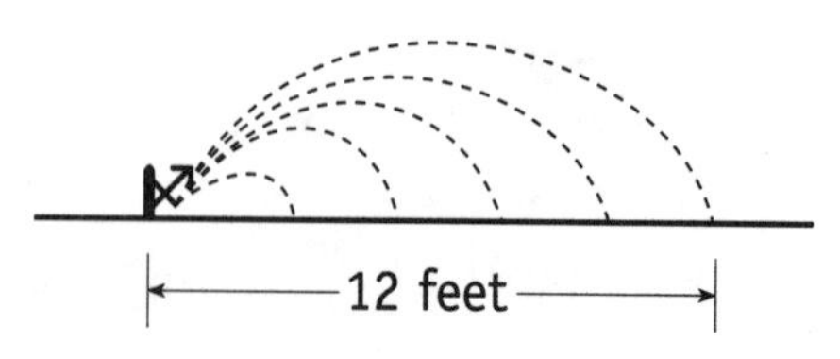

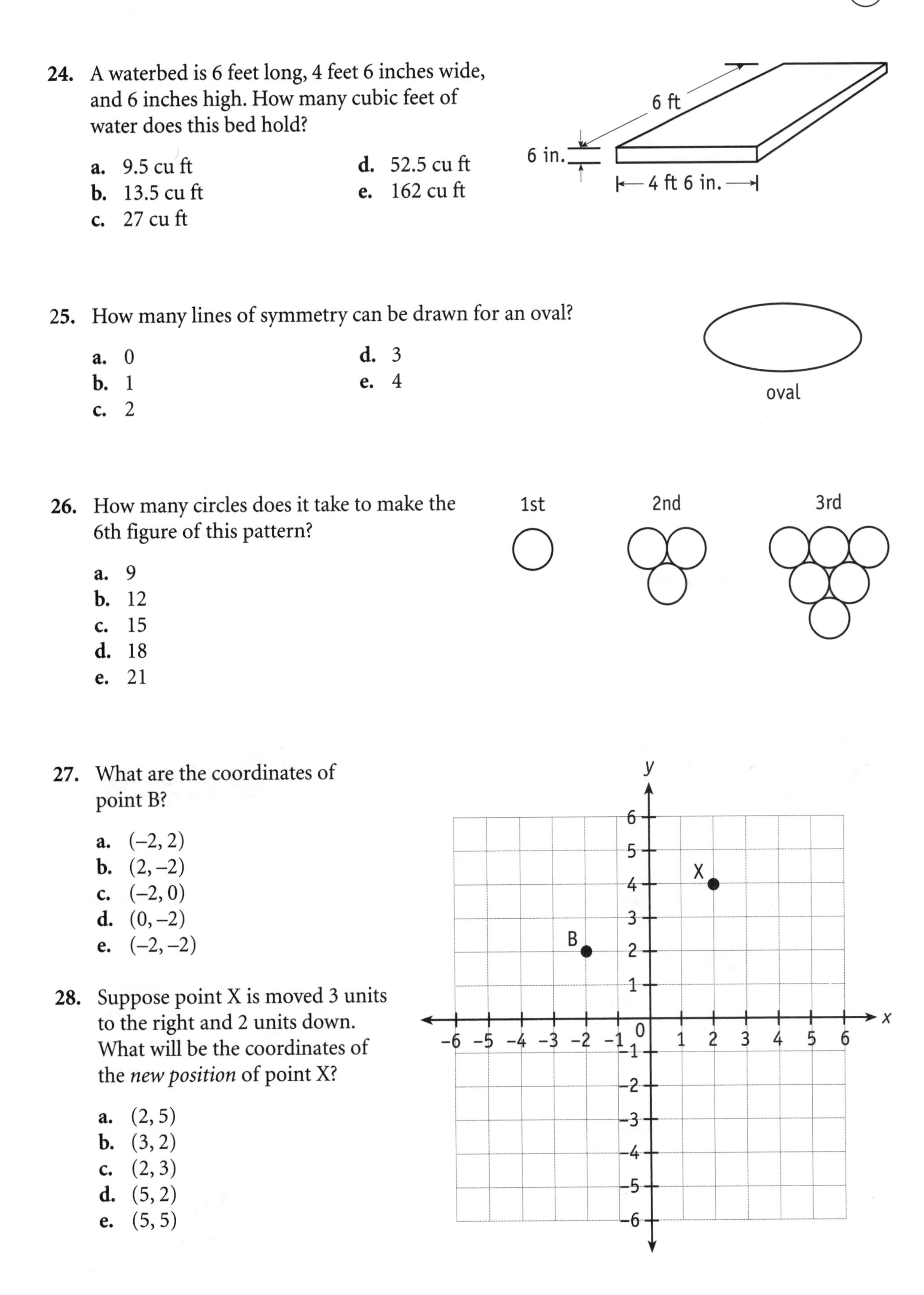

24. A waterbed is 6 feet long, 4 feet 6 inches wide, and 6 inches high. How many cubic feet of water does this bed hold?

a. 9.5 cu ft
b. 13.5 cu ft
c. 27 cu ft
d. 52.5 cu ft
e. 162 cu ft

25. How many lines of symmetry can be drawn for an oval?

a. 0
b. 1
c. 2
d. 3
e. 4

26. How many circles does it take to make the 6th figure of this pattern?

a. 9
b. 12
c. 15
d. 18
e. 21

27. What are the coordinates of point B?

a. (–2, 2)
b. (2, –2)
c. (–2, 0)
d. (0, –2)
e. (–2, –2)

28. Suppose point X is moved 3 units to the right and 2 units down. What will be the coordinates of the *new position* of point X?

a. (2, 5)
b. (3, 2)
c. (2, 3)
d. (5, 2)
e. (5, 5)

29. The sales tax in Kristen's state is 6%. Which algebraic expression gives the sales tax on a sweater that costs n dollars?

a. $n - 0.06$ **b.** $n + 0.06$ **c.** $(0.06)n$ **d.** $0.06 \div n$ **e.** $n \div 0.06$

30. What is the value of the algebraic expression below when $a = 4$, $b = 1$, and $c = 3$?

$2a^2 + b^2 - 2ac$

a. 7 **b.** 9 **c.** 12 **d.** 15 **e.** 18

31. What is the solution to the equation $3y = 24$?

a. $y = 4$ **b.** $y = 6$ **c.** $y = 8$ **d.** $y = 12$ **e.** $y = 24$

32. What is the solution to the equation $\frac{x}{4} = 8$?

a. $x = 32$ **b.** $x = 28$ **c.** $x = 12$ **d.** $x = 4$ **e.** $x = 2$

Pre-Algebra Posttest B Chart

If you missed more than one problem on any group below, review the practice pages for those problems. Then redo the problems you got wrong before going on to Using Number Power. If you had a passing score, redo any problems you missed and go on to Using Number Power on page 177.

PROBLEM NUMBERS	SKILL AREA	PRACTICE PAGES
1, 2, 3, 4	number skills	10–29
5, 6, 7, 8	word problem skills	32–51
9, 10, 11, 12	problem-solving strategies	54–67
13, 14, 15, 16	data analysis	70–85
17, 18, 19, 20	probability	88–103
21, 22, 23, 24	geometry	106–121
25, 26, 27, 28	spatial sense and patterns	124–137
29, 30, 31, 32	algebra	140–159

Using Number Power

Prime Numbers and Composite Numbers

A **factor** is a number that divides evenly into another number. Every whole number has at least two factors: itself and the number 1. Most whole numbers have more than two factors. For example, the number 8 has four factors: 1, 2, 4, and 8.

- A whole number that has only two factors (itself and 1) is called a **prime number.**
  Examples of prime number: 2, 3, 5, and 7
- A whole number that has more than two factors is called a **composite number.**
  Examples of composite numbers: 4, 6, 8, and 9

Prime Numbers	Factors (two only)
2	1, 2
3	1, 3
5	1, 4
7	1, 7

Composite Numbers	Factors (more than two)
4	1, 2, 4
6	1, 2, 3, 6
8	1, 2, 4, 8
9	1, 3, 9

The number 1 is neither a prime number nor a composite number.

- Every whole number can be written as a product of two or more of its factors.

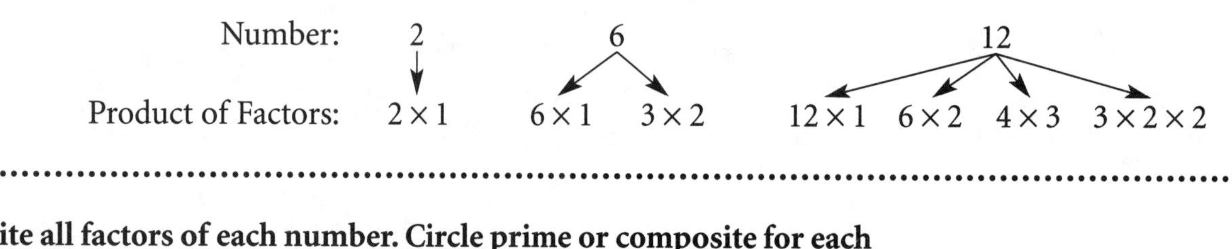

Write all factors of each number. Circle prime or composite for each number.

1. 13
   prime composite
2. 15
   prime composite
3. 27
   prime composite
4. 29
   prime composite
5. 30
   prime composite
6. 48
   prime composite

Write each missing factor.

7. $3 = \_\_\_\_ \times 1$ $4 = \_\_\_\_ \times 2$ $9 = \_\_\_\_ \times 3$ $12 = \_\_\_\_ \times 3$

8. $20 = \_\_\_\_ \times 4$ $56 = \_\_\_\_ \times 8$ $70 = \_\_\_\_ \times 7$ $44 = \_\_\_\_ \times 11$

Each number is written as a product of three factors. Write the two missing factors, neither of which is the number 1.

9. $12 = \_\_\_\_ \times \_\_\_\_ \times 3$ $16 = \_\_\_\_ \times \_\_\_\_ \times 2$ $28 = \_\_\_\_ \times \_\_\_\_ \times 7$

10. $36 = \_\_\_\_ \times \_\_\_\_ \times 4$ $50 = \_\_\_\_ \times \_\_\_\_ \times 5$ $72 = \_\_\_\_ \times \_\_\_\_ \times 4$

Circle each *prime number* between 1 and 35.

11. 1 2 3 4 5 6 7 8 9 10 11 12 13 14 15 16 17 18 19 20

21 22 23 24 25 26 27 28 29 30 31 32 33 34 35

Prime-Factorization Form

Every composite number can be written as a product of **prime factors** (factors that are prime numbers). This is called **prime-factorization** form.

Number	Product of Prime Factors (prime-factorization form)	Written Using Exponents
8	$2 \times 2 \times 2$	2^3
20	$2 \times 2 \times 5$	$2^2 \cdot 5$
72	$2 \times 2 \times 2 \times 3 \times 3$	$2^3 \cdot 3^2$
110	$2 \times 5 \times 11$	$2 \cdot 5 \cdot 11$
125	$5 \times 5 \times 5$	5^3

Write each number in prime-factorization form.

12. 6 = 9 = 10 =

13. 12 = 24 = 32 =

14. 49 = 50 = 86 =

Reading an Inch Ruler

In the U.S. customary system, short distances are measured with an inch ruler. Each inch (in.) is divided into fractions, the smallest unit of the ruler being $\frac{1}{16}$ inch.

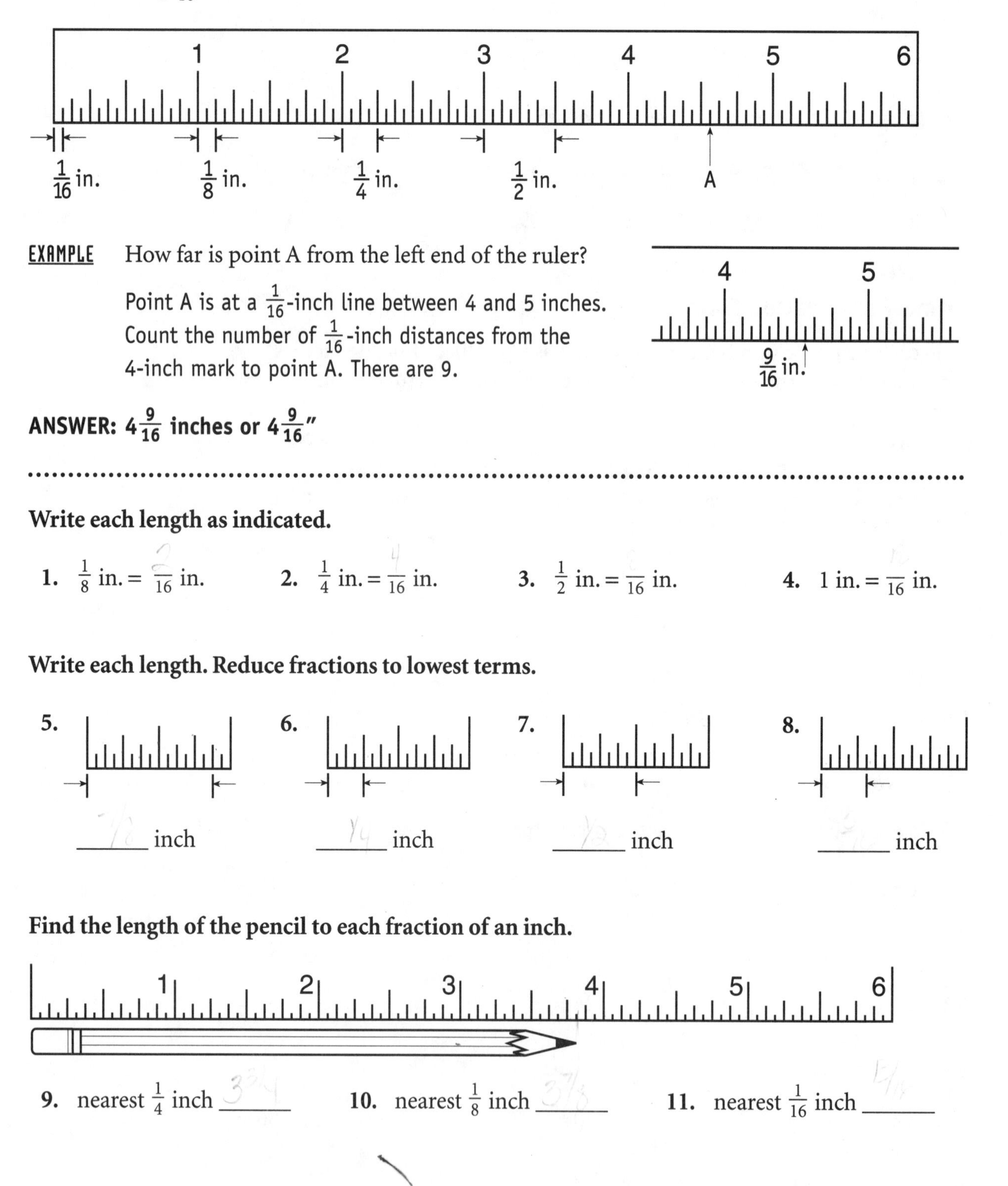

EXAMPLE How far is point A from the left end of the ruler?

Point A is at a $\frac{1}{16}$-inch line between 4 and 5 inches. Count the number of $\frac{1}{16}$-inch distances from the 4-inch mark to point A. There are 9.

ANSWER: $4\frac{9}{16}$ inches or $4\frac{9}{16}$"

Write each length as indicated.

1. $\frac{1}{8}$ in. = $\frac{\ }{16}$ in.
2. $\frac{1}{4}$ in. = $\frac{\ }{16}$ in.
3. $\frac{1}{2}$ in. = $\frac{\ }{16}$ in.
4. 1 in. = $\frac{\ }{16}$ in.

Write each length. Reduce fractions to lowest terms.

5. ______ inch
6. ______ inch
7. ______ inch
8. ______ inch

Find the length of the pencil to each fraction of an inch.

9. nearest $\frac{1}{4}$ inch ______
10. nearest $\frac{1}{8}$ inch ______
11. nearest $\frac{1}{16}$ inch ______

Reading a Centimeter Ruler

In the metric system, short distances are measured on a centimeter ruler. Each centimeter (cm) is divided into 10 millimeters (mm). Pictured below is a 15-centimeter ruler.

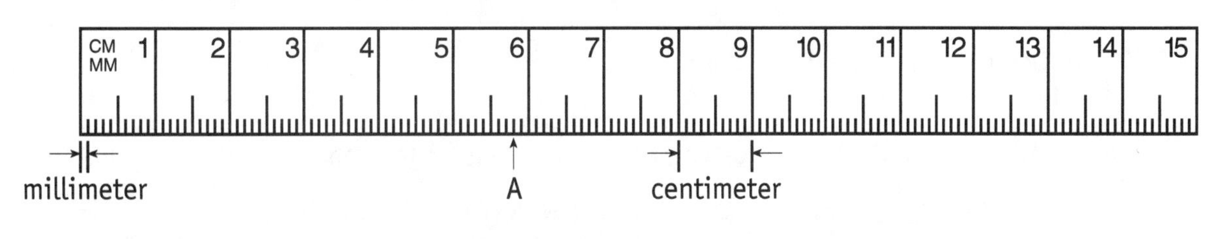

Point A is 5 centimeters and 8 millimeters (5 cm 8 mm) from the left end of the ruler. Because 1 cm = 10 mm, you also can write 5 cm 8 mm as 5.8 cm *or* as 58 mm.

EXAMPLE 1 Write 3 cm 6 mm as centimeters.

ANSWER: 3 cm 6 mm = 3.6 cm

EXAMPLE 2 Write 6.2 centimeters as millimeters.

ANSWER: 6.2 cm = 62 mm

EXAMPLE 3 Write 119 millimeters in two other ways.

ANSWER: 119 mm = 11 cm 9 mm *or* 11.9 cm

Write each length as indicated.

1. Write 9 cm 2 mm as centimeters.
2. Write 5.4 cm as millimeters.
3. Write 136 millimeters in two other ways.

Write each length pictured below.

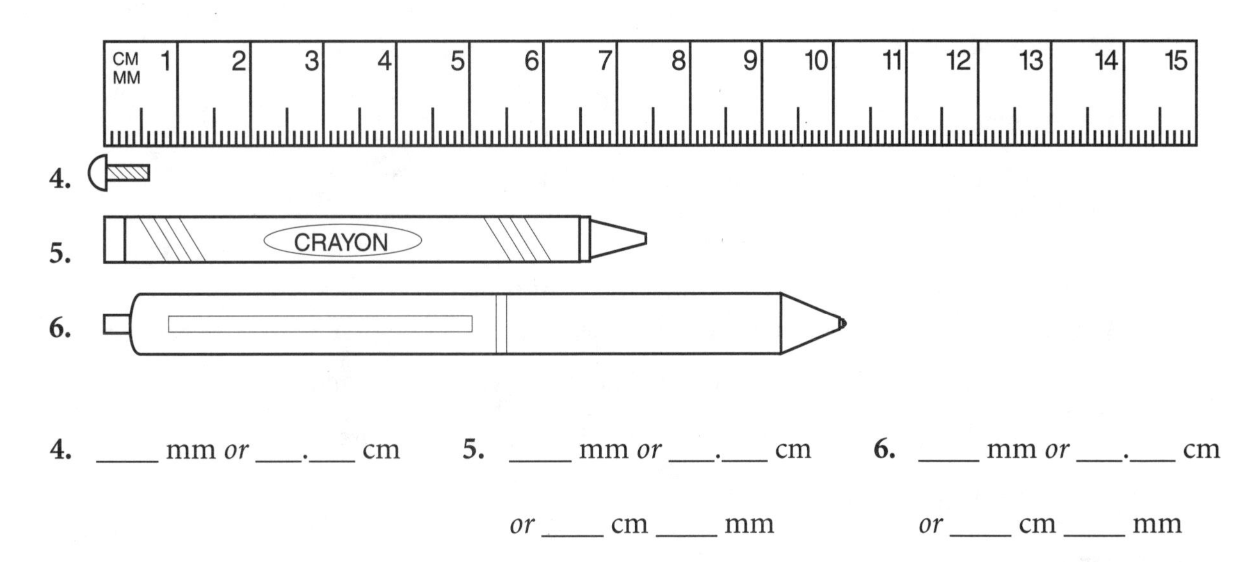

4. _____ mm *or* ____.____ cm
5. _____ mm *or* ____.____ cm *or* _____ cm _____ mm
6. _____ mm *or* ____.____ cm *or* _____ cm _____ mm

Distance Formula

The distance formula $d = rt$ tells how to find the distance traveled (d) when you know the rate (r) and the time of travel (t). *Rate* is another word for *speed.*

EXAMPLE 1 What distance can Jandi drive in 6 hours if she averages 55 miles per hour?

STEP 1 Identify r and t.

$r = 55$ miles per hour

$t = 6$ hours

STEP 2 Substitute values for r and t.

$d = rt = (55)(6) =$ **330 miles**

ANSWER: 330 miles

The distance formula can be rewritten in two other ways.

- as a rate formula
- as a time formula

As a **rate formula,** $r = \frac{d}{t}$.

The rate formula is used to find the rate (speed) when the distance and time are known.

EXAMPLE 2 On part of her trip, Jandi drives 424 miles in 8 hours. What is her average speed?

STEP 1 Identify d and t.

$d = 424$ miles

$t = 8$ hours

STEP 2 Substitute values for d and t.

$r = \frac{d}{t} = \frac{424}{8} =$ **53 miles per hour**

ANSWER: 53 miles per hour

As a **time formula,** $t = \frac{d}{r}$.

The time formula is used to find the time when the distance and rate (speed) are known.

EXAMPLE 3 On the final day, Jandi drives 345 miles. If she averages 46 miles per hour, how long will the drive take?

STEP 1 Identify d and r.

$d = 345$ miles

$r = 46$ miles per hour

STEP 2 Substitute values for d and r.

$t = \frac{d}{r} = \frac{345}{46} =$ **7.5 hours**

ANSWER: 7.5 hours (7 hours 30 minutes)

In each problem below, decide whether you are asked to find the distance (*d*), rate (*r*), or time (*t*). Then solve the problem.

1. How far can Stacey drive in 5 hours if she averages 60 miles per hour?

2. If Brandon drives 280 miles in 5 hours, what is his average speed per hour?

3. Brent is taking a bus to Chicago, 240 miles from his home. If the bus averages 60 miles per hour, how long will the trip take?

4. On a trip to Portland, Dolores averaged 60 miles per hour while driving for 6 hours 30 minutes. How far did Dolores drive during this time?

5. Caren hiked 8 miles in 10 hours. At this rate, how many miles did Caren hike each hour?

6. If Zander can ride his bike at an average speed of 15 miles per hour, how long will it take him to ride 40 miles?

7. At the Aquatic Center, Isaac can swim 1.5 laps in 1 minute. How many laps can Isaac swim in 40 minutes?

8. Marsha walked from her house to the library, a distance of 2.4 miles. If she walks at the rate of 3 miles per hour, how long did the walk take? Express your answer in minutes.

Using a Protractor

A **protractor** is a tool used to measure and to draw angles. A protractor is shaped like half of a circle and can measure angles up to 180°. Carpenters, machinists, contractors, and others who read blueprints use protractors in their work.

Most protractors have two scales, one running clockwise (↷), and the other running counterclockwise (↶).

Measuring an Angle

EXAMPLE 1 To measure an angle with a protractor

- Place the 0° baseline of the protractor over one side of the angle.
- Place the center mark of the protractor over the angle vertex.
- Read the point where the second side of the angle crosses the protractor scale.

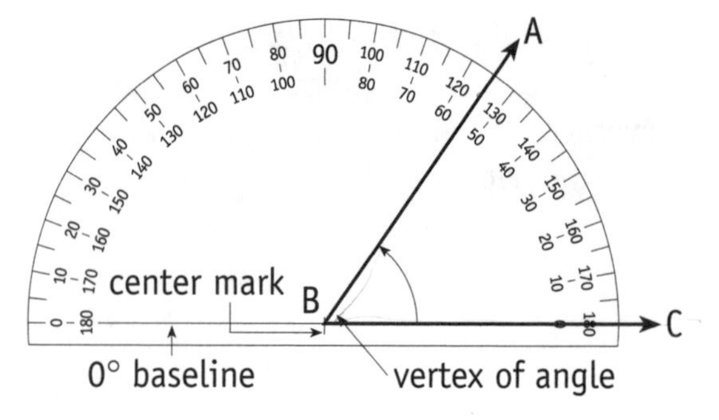

ANSWER: $\angle ABC = 55°$

Find the measure of each angle.

1. $\angle XYZ =$ _____

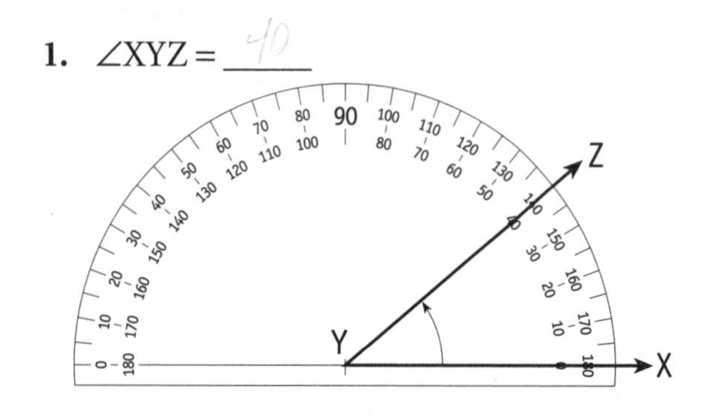

2. $\angle LMN =$ _____

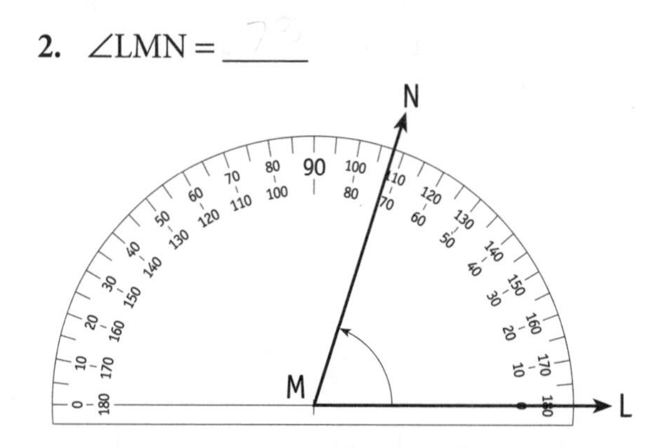

3. $\angle EFG =$ _____

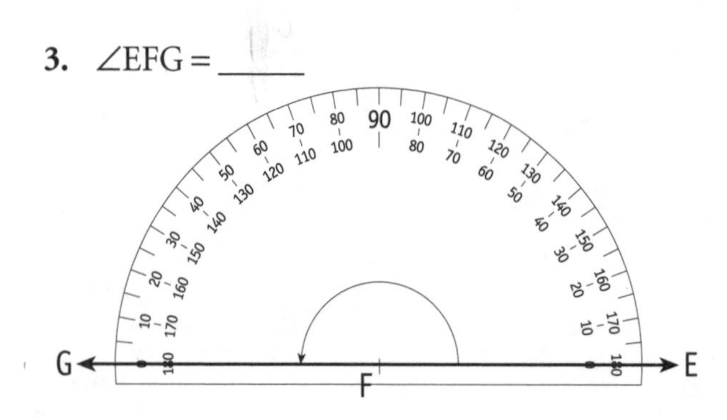

4. $\angle RST =$ _____

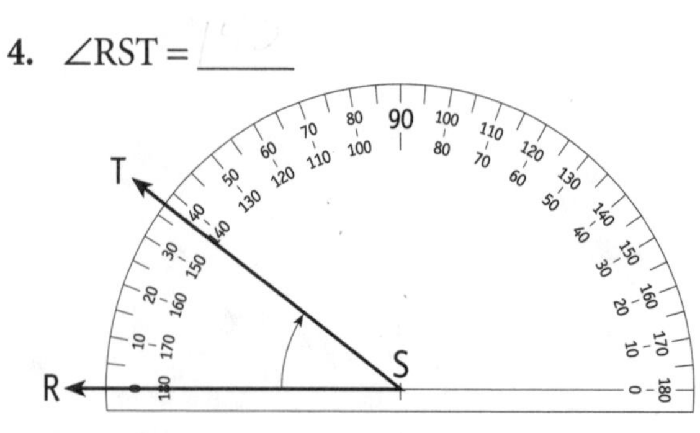

Drawing an Angle

EXAMPLE 2 To draw an angle with a protractor

- Draw one side along the 0° baseline.
- Place the protractor over the vertex of the angle you are drawing.
- Draw the second side to cross the protractor scale at the correct measure.

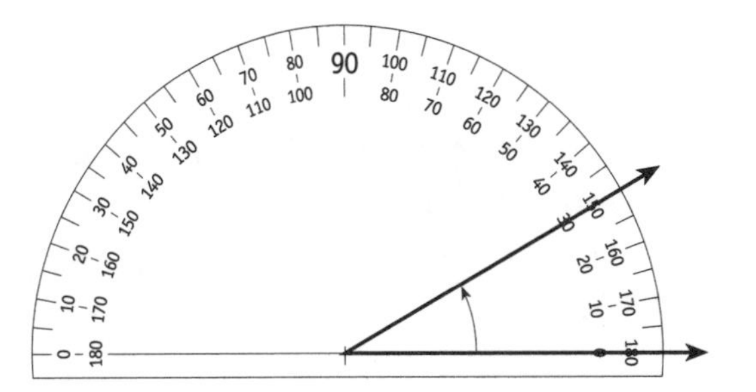

The angle at the right is a 30° angle, drawn along the counterclockwise scale.

Draw each angle on the protractor.

5. Draw an 80° angle that opens counterclockwise.

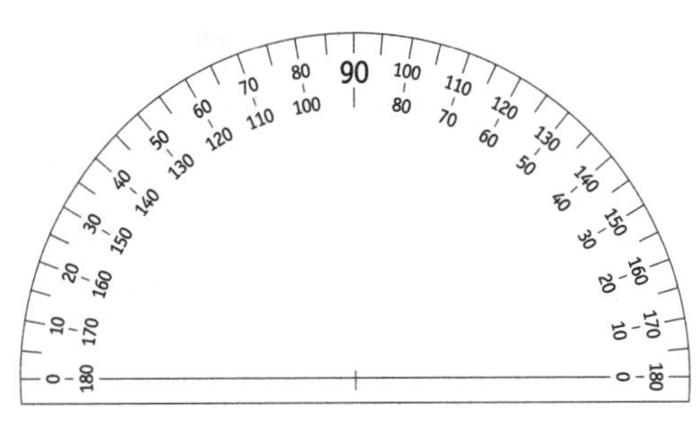

6. Draw a 142° angle that opens counterclockwise.

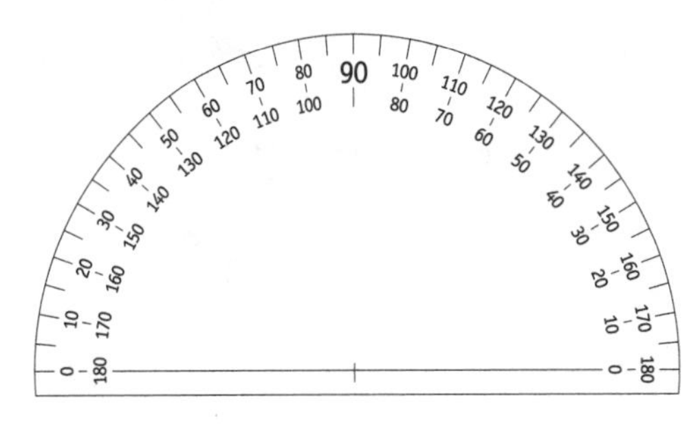

7. Draw a 90° angle that opens clockwise.

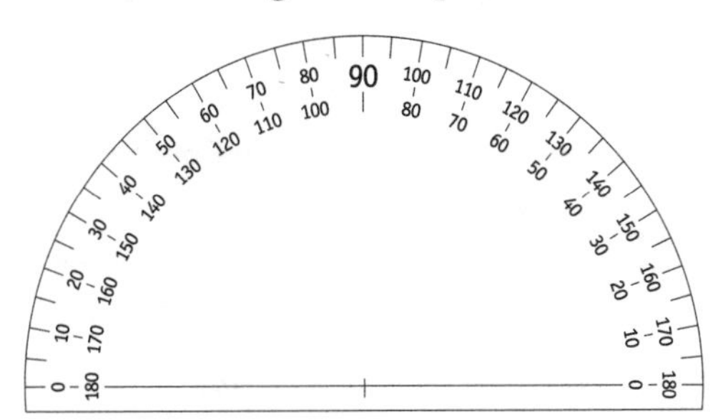

8. Draw a 136° angle that opens clockwise.

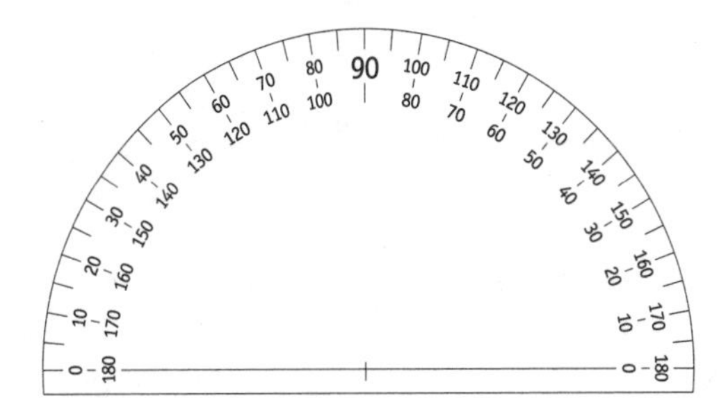

Surface Area

A 3-dimensional object, such as a box (rectangular solid), has both volume and surface area. The **surface area** of a box is the sum of the areas of the six outer surfaces. The box pictured below is 12 inches long, 3 inches wide, and 6 inches high. The box has six surfaces. These surfaces are best seen if the box is drawn unfolded.

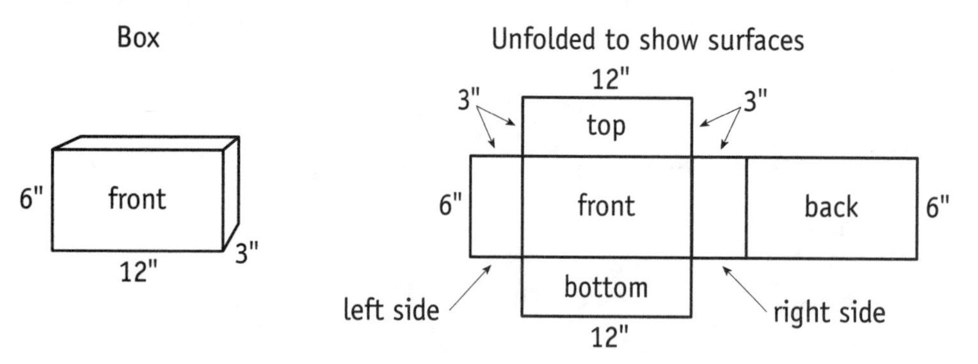

The surface area is the sum of the areas of the six surfaces.

Surface	Area (sq in.)
left side:	$6 \times 3 = 18$
right side:	$6 \times 3 = 18$
front:	$12 \times 6 = 72$
back:	$12 \times 6 = 72$
top:	$12 \times 3 = 36$
bottom:	$12 \times 3 = 36$

Math Tip
The formula for the surface area of a cube is
surface area $= 6s^2$.
Can you explain why?

ANSWER: Total surface area = 252 sq in.

Solve each problem.

A box measures 3 feet long, 2 feet wide, and $1\frac{1}{2}$ feet high.

1. **a.** Label each surface with its dimensions.

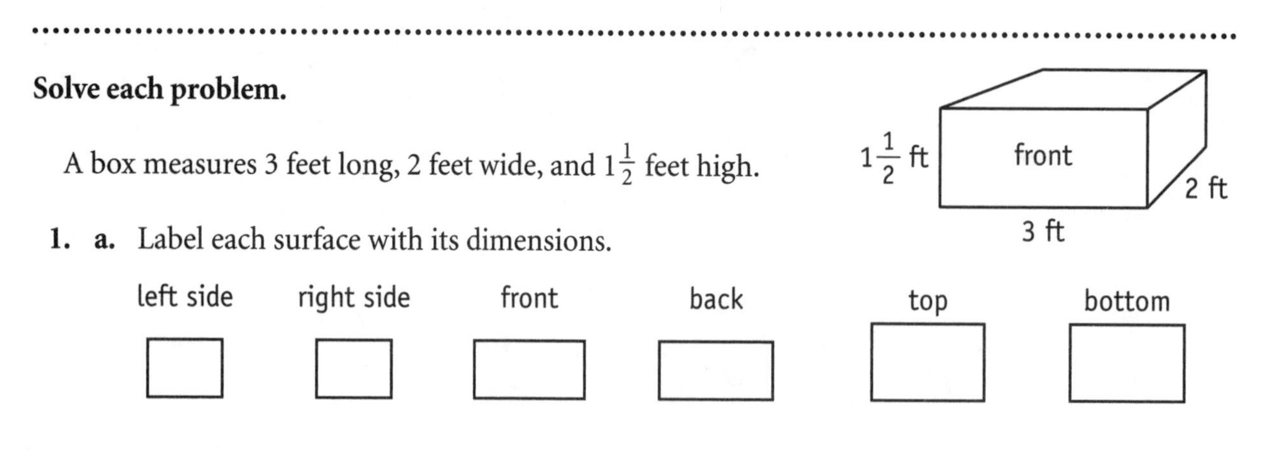

b. What is the surface area of the box?

2. **a.** What is the volume of the brick shown here?

b. What is the surface area of the brick?

3. **a.** What is the volume of the suitcase pictured at the right? Express your answer in cubic feet. (**Hint:** 8 in. = $\frac{2}{3}$ ft)

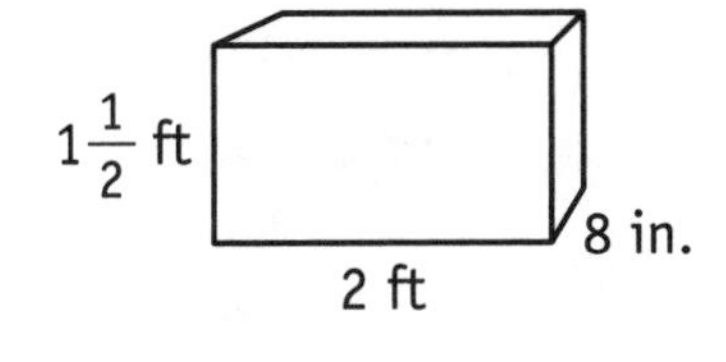

b. What is the surface area of the suitcase? Express your answer in square feet.

4. The pyramid at the right has a square base and four triangular sides. Counting the base, what is the surface area of this pyramid? (area of a triangle = $\frac{1}{2}bh$)

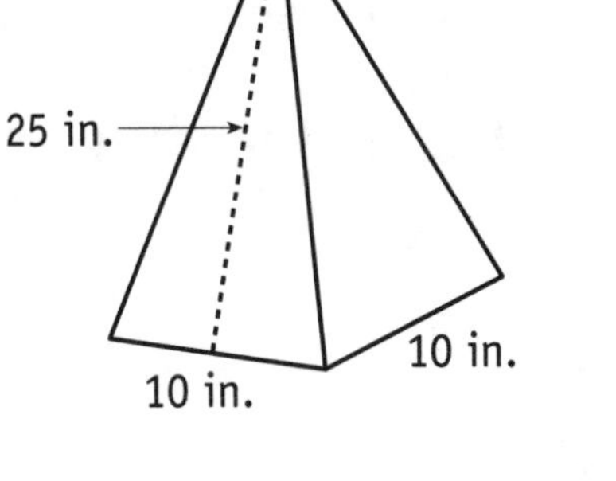

5. A cylindrical tank is 6 feet high. The *diameter* of the tank is 4 feet. To the nearest square foot, what is the surface area of the tank? (**Hint:** The length of the rectangle is the circumference of the circle.)

Unfolded, a cylinder can be pictured as a rectangle and two circles.

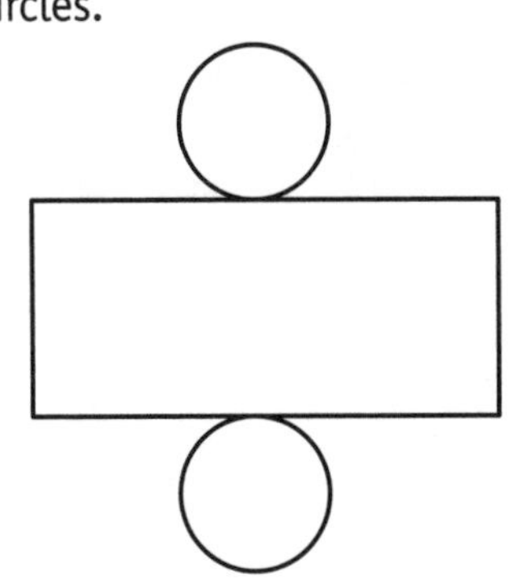

Similar Figures

Proportions are often used to solve problems involving similar figures. Two figures are similar if they have the same shape (angles of equal measure and proportional dimensions).

Finding a Missing Dimension

EXAMPLE 1 Tony wants to enlarge a photograph that measures 5 inches wide by 3 inches high so that the larger photograph measures 8 inches wide. What will be the height of the enlargement?

STEP 1 Let h be the unknown height. Write a proportion as follows.

$$\frac{\text{height of enlargement}}{\text{height of original}} = \frac{\text{width of enlargement}}{\text{width of original}}$$

$$\frac{h}{3} = \frac{8}{5}$$

STEP 2 Solve for h.

$$h = \frac{8 \times 3}{5} = \frac{24}{5} = 4.8 \text{ inches}$$

ANSWER: 4.8 inches

Enlargement

h

8 in.

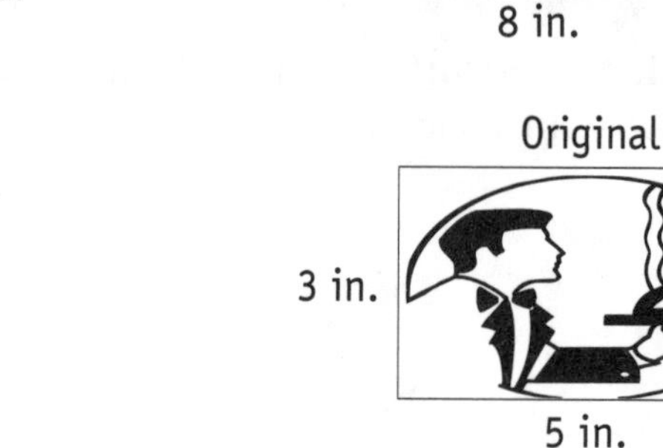

Use proportions to solve each problem.

1. As part of a class project, Keisha is having a large poster made from a photograph. If the height of the poster will be 33 inches, what will be its width?

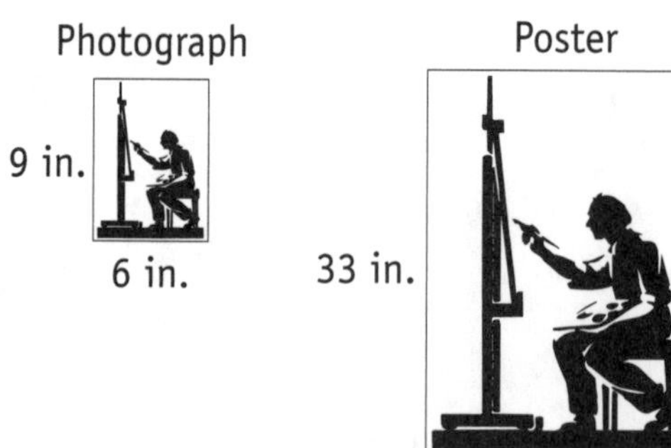

2. Scott wants to have a wallet-size photograph made from a larger photo. If the width of the wallet-size copy is 2 inches, what will be its height?

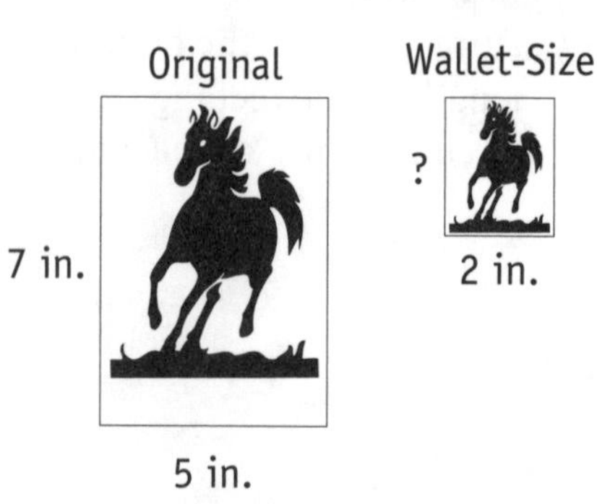

3. Two trees stand side by side. The short tree is 7 feet high and casts a shadow of 4 feet. At the same time of day, the tall tree casts a shadow of 11 feet. What is the height of the tall tree?

$$\frac{\text{height of tall tree}}{\text{height of short tree}} = \frac{\text{length of long shadow}}{\text{length of short shadow}}$$

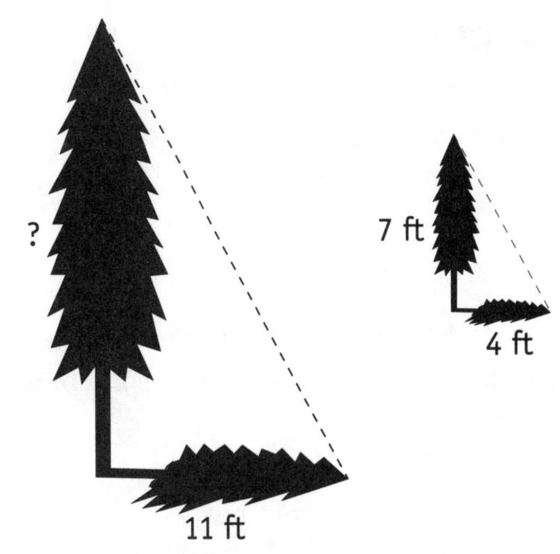

Similar right triangles are formed by the trees and their shadows. The hypotenuse of each triangle is formed by the top edge of the shadow and the top of the tree.

Working with a Scale Drawing

Scale drawings, such as blueprints and maps, are also examples of the use of similar figures. In scale drawings, pictured objects are proportional to the objects they represent. A **scale** tells the ratio of sizes. For example, *Scale: 1 inch = 20 inches* means that the ratio of each dimension of the actual object to the dimension of the drawn object is 20 to 1.

EXAMPLE 2 On a blueprint for a deck, the scale reads "1 inch = 24 inches." What is the actual length of a bench that measures 3.5 inches on the blueprint?

To find the length, multiply 24 by 3.5.

bench length = $24 \times 3.5 =$ **84 inches**

ANSWER: 84 inches

EXAMPLE 3 A map of the United States has a scale "1 inch = 200 miles." What is the distance between two cities whose map distance is 7 inches?

To find the distance, multiply 200 by 7.

distance = $200 \times 7 =$ **1,400 miles**

ANSWER: 1,400 miles

Solve each problem.

4. On a blueprint for a house, the scale reads "1 inch = 18 inches." What is the actual length of a hallway that measures $8\frac{1}{2}$ inches on the blueprint?

5. On a map of New York, the scale reads "1 inch = 20 miles." What is the air distance between two cities whose map distance is $4\frac{1}{4}$ inches?

Perspective

Perspective is the skill of visualizing a figure from different points of view. Perspective is used in fields such as graphic arts and building design.

For problems 1–3, refer to the drawings below. The drawings are the front view and right-side view of a figure that is made from six cubes.

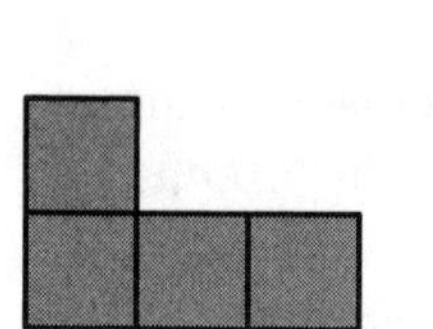

Front View

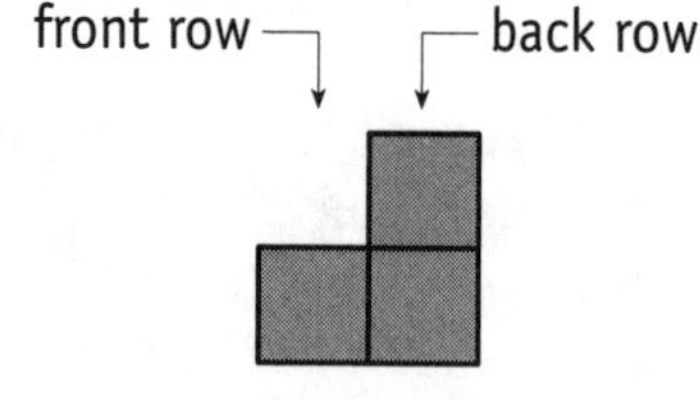

Right-side View

You may find it helpful to use sugar cubes or other objects to discover what figures give the same front and right-side views as those shown.

1. The cubes are placed in two rows.

   a. What is the *greatest number* of cubes that may be in the front row?

   b. What is the *least number* of cubes that may be in the front row?

2. On the grids below, draw the back view and the left-side view of the figure.

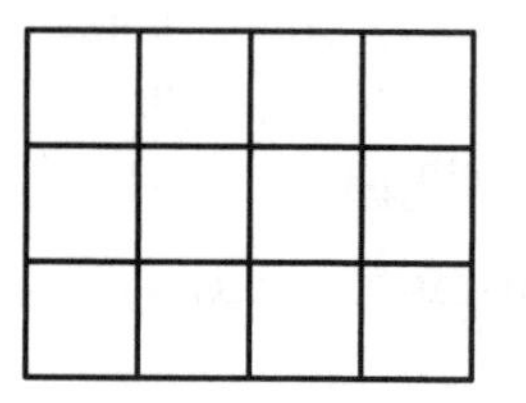

Back View

Left-side View

3. On the grids below, draw two of the four possible top views of the figure.

Top View #1

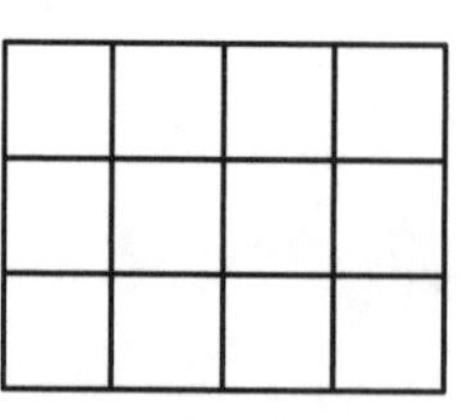

Top View #2

4. Imagine you can cut out the figures below and fold them along the lines. Circle each figure that can be folded into a cube.

Cube

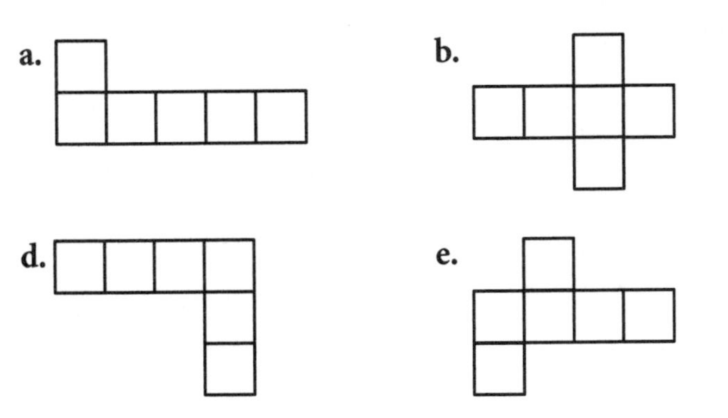

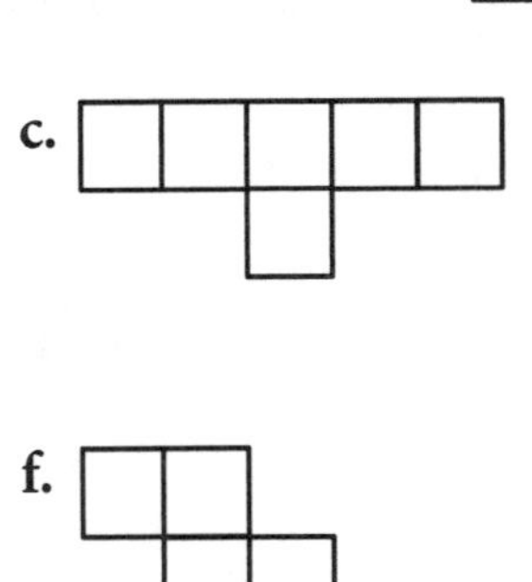

5. Five different shapes can be formed by joining four squares along their sides. Two shapes are not different if they are congruent. (One can be placed exactly over the other.) One shape is drawn below. Draw the other four shapes.

6. The figure at the right is made out of seven cubes. On the grids below, draw each view as indicated.

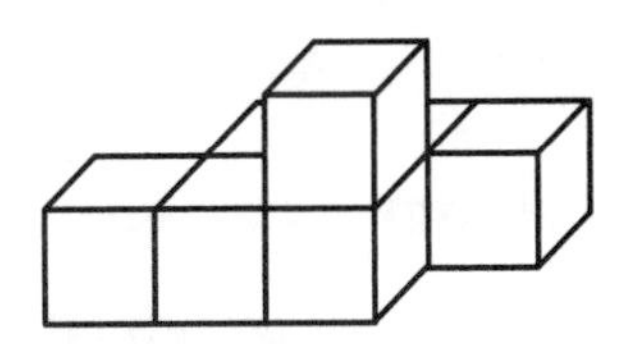

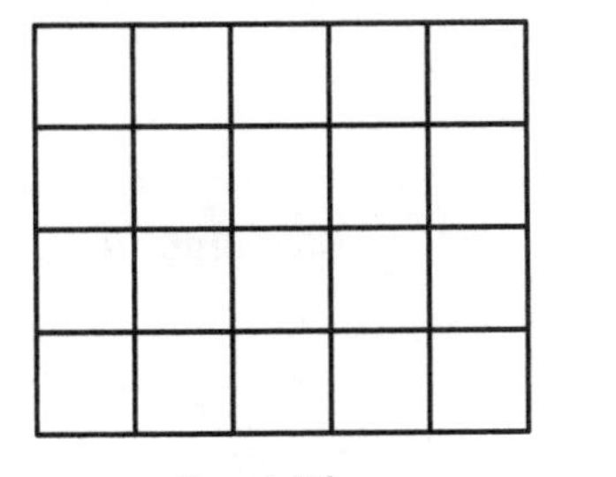

Front View

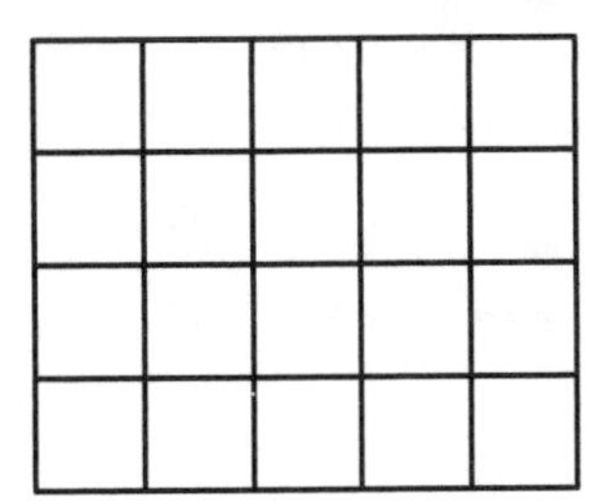

Top View

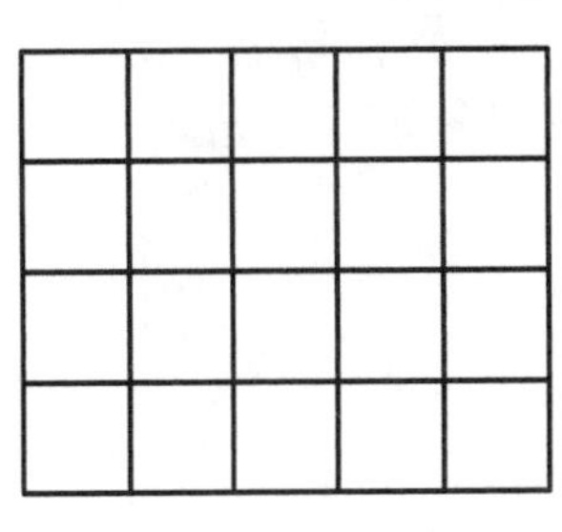

Left-side View

Regular Polygons and Symmetry

In a **regular polygon,** all sides have equal length and all angles have equal measure. A regular polygon has two kinds of symmetry.

- **Reflection symmetry**—symmetry across one or more lines of symmetry
- **Rotation symmetry**—symmetry arising from rotation around the center point of a figure

Three regular polygons are shown below.

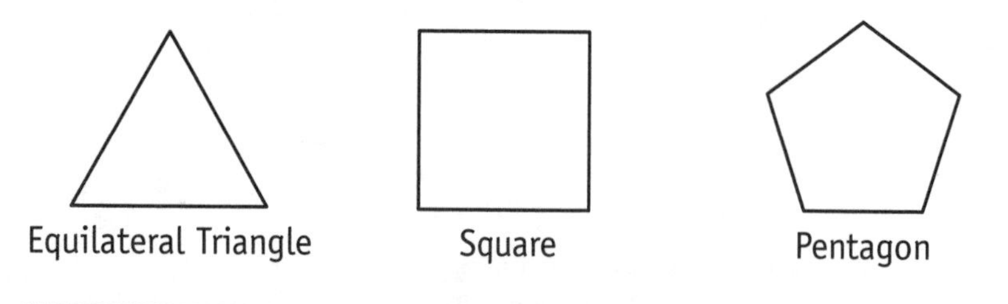

Reflection Symmetry and Regular Polygons

1. **a.** Draw all lines of symmetry for each regular polygon above.

   b. How many lines of symmetry does each figure have?

   equilateral triangle: ______ square: ______ pentagon: ______

2. How many lines of symmetry does a 10-sided regular polygon have?

Rotation Symmetry and Regular Polygons

3. Suppose you rotate each regular polygon around its center. What is the *minimum angle* you must rotate each figure before it fits exactly on top of its original shape?

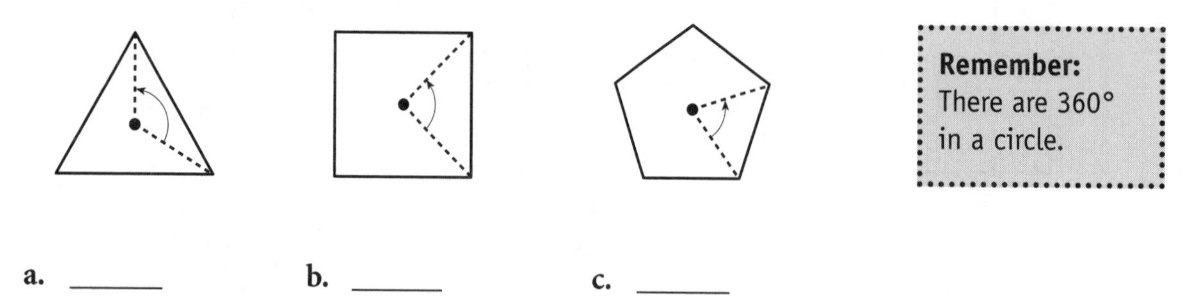

a. ______ **b.** ______ **c.** ______

4. How many degrees must you rotate a 10-sided regular polygon before it fits exactly on top of its original shape?

Figurate Numbers

Ancient Greeks studied **figurate numbers,** numbers associated with geometric figures. **Triangular numbers** are associated with equilateral triangles. The first four triangular figures and triangular numbers are shown below. The first figure is a single dot.

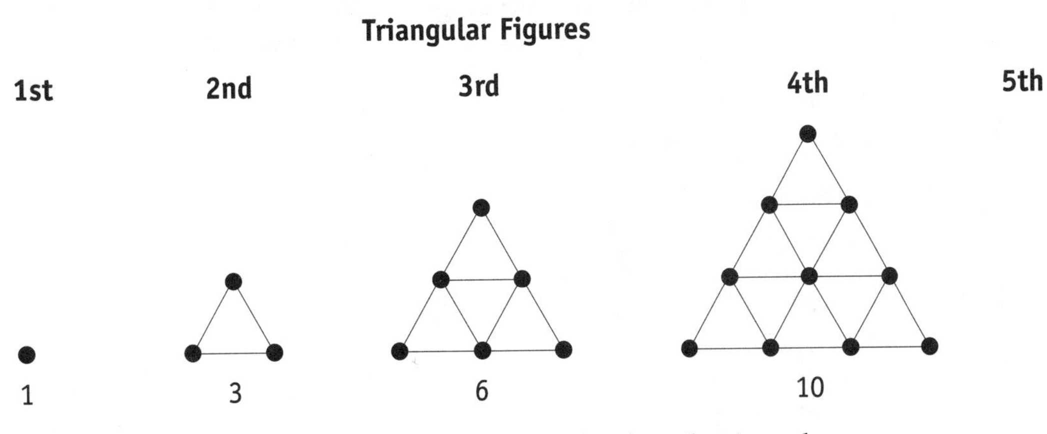

1. Draw the 5th triangular figure above. Write the 5th triangular number below your drawing.

2. From the pattern above, determine the 6th and 7th triangular numbers. _____ 6th _____ 7th

Square numbers are associated with squares. The first three square figures and square numbers are shown below. Again, the first figure is a single dot.

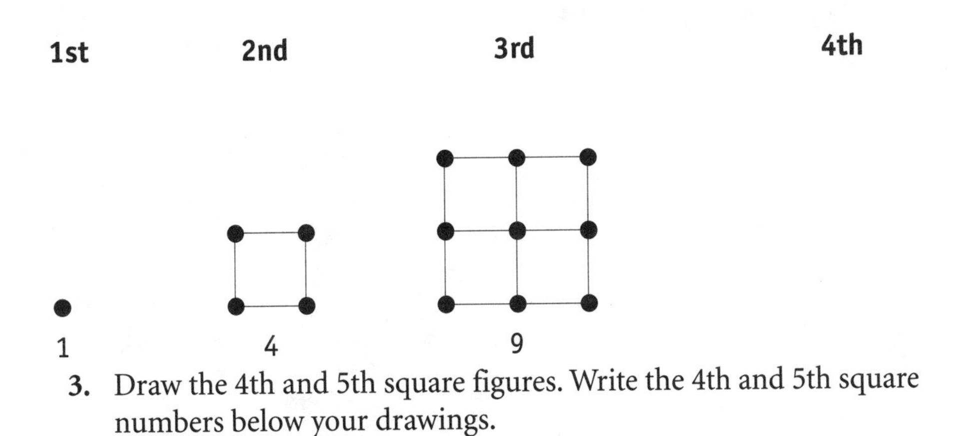

3. Draw the 4th and 5th square figures. Write the 4th and 5th square numbers below your drawings.

4. From the pattern above, determine the 6th and 7th square numbers. _____ 6th _____ 7th

Pythagorean Theorem

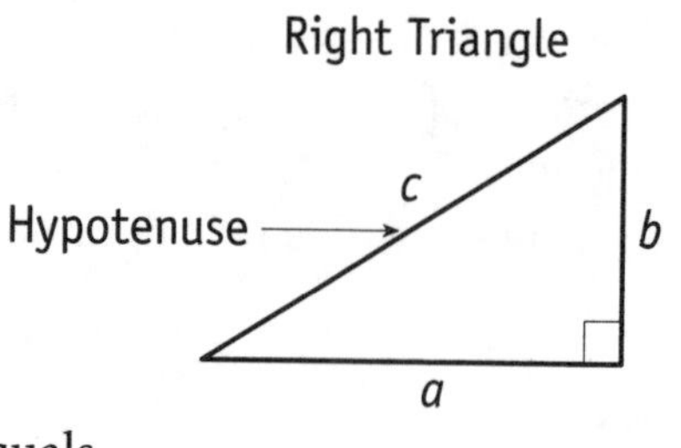

In a right triangle, the longest side—the side opposite the 90° angle—is called the **hypotenuse.**

Pythagoras, a famous Greek mathematician, discovered an important relationship between the hypotenuse of a right triangle and the two shorter sides. This relationship is called the **Pythagorean theorem.**

In words: In a right triangle, the square of the hypotenuse equals the sum of the squares of the two shorter sides.

In symbols: $c^2 = a^2 + b^2$ (using the side names on the triangle above)

Finding the Length of the Hypotenuse

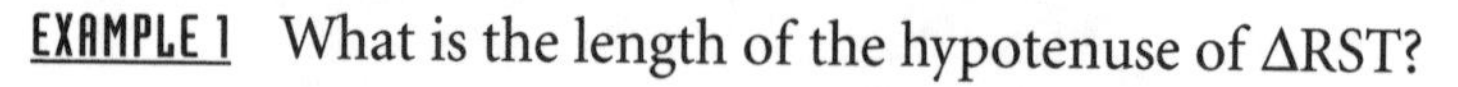

EXAMPLE 1 What is the length of the hypotenuse of ΔRST?

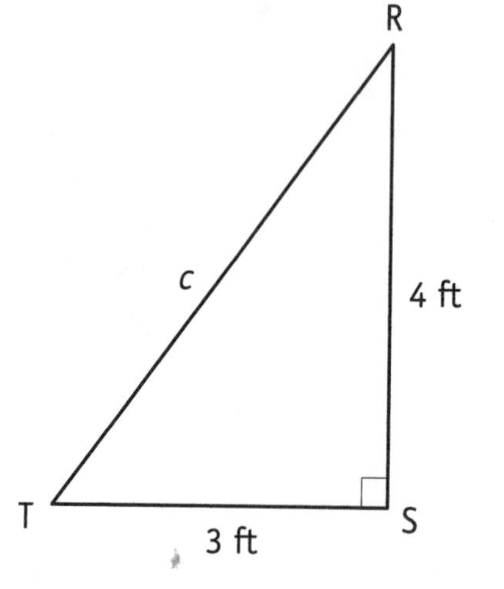

STEP 1 Square each shorter side.

$a^2 = 3^2 = 9 \qquad b^2 = 4^2 = 16$

STEP 2 Add the squares found in Step 1.

$$\begin{aligned} c^2 &= a^2 + b^2 \\ &= 9 + 16 \\ &= 25 \end{aligned}$$

STEP 3 To find c, take the square root of 25.

$c = \sqrt{25} = \mathbf{5}$

ANSWER: The length of the hypotenuse is 5 feet.

Find the length of the hypotenuse in each triangle below.

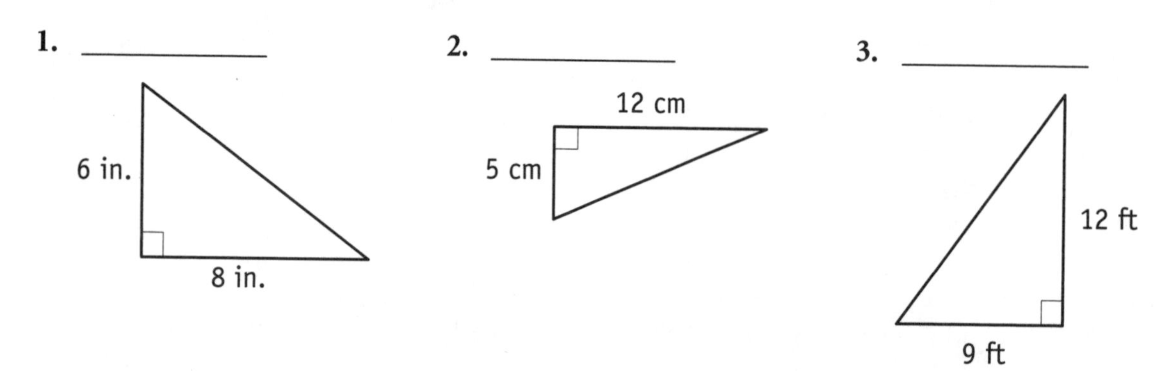

Finding the Length of a Shorter Side

You can find the length of a shorter side if you know the lengths of the hypotenuse and the other side.

In words: Side a squared equals hypotenuse squared minus side b squared.

In symbols: $b^2 = c^2 - a^2$ (This is another way to write the Pythagorean theorem.)

EXAMPLE 2 In ΔXYZ, the hypotenuse is 5 feet and one side is 3 feet. Find the length of the other side.

To find the length of the other side, subtract squares.

$$b^2 = c^2 - a^2$$
$$= 5^2 - 3^2$$
$$= 25 - 9 = 16$$
$$b = \sqrt{16} = \mathbf{4}$$

X
5 ft
?
Z
Y
3 ft

ANSWER: The length of the side is **4 feet.**

Find the length of the unlabeled side in each triangle below. The length of each side is a whole number.

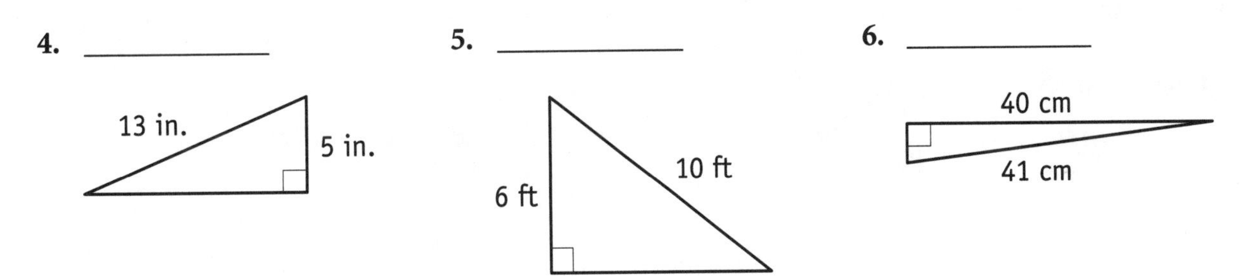

Use the square root key on your calculator ($\sqrt{x}$) to find each distance to the nearest whole number.

7. the distance between home and 2nd base

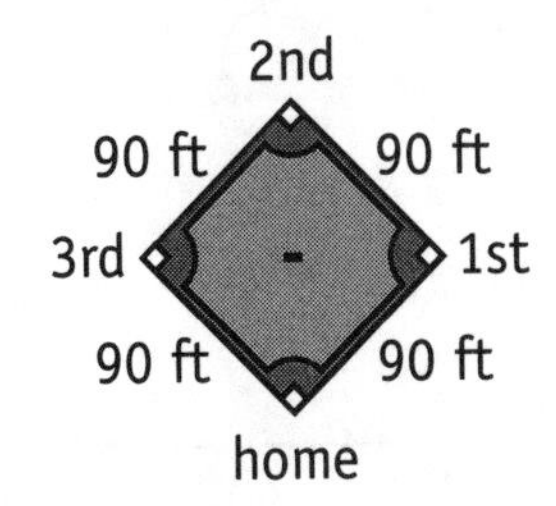

8. the distance across the rectangular garden along the dotted line

?
6 yd
16 yd

9. the distance of the ship from the shore

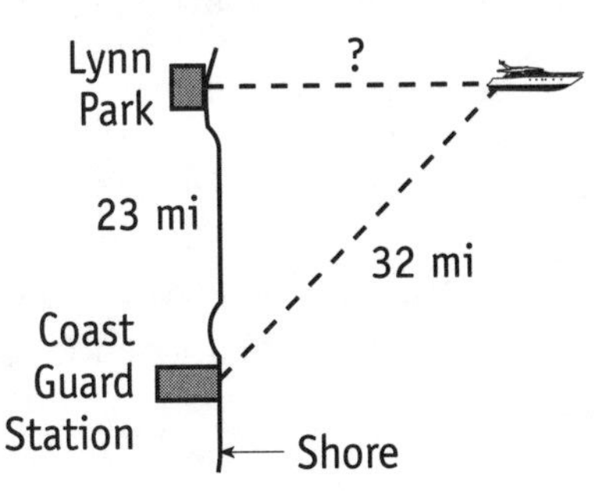

Fahrenheit and Celsius Temperatures

Temperature is measured in units called **degrees.** Both Fahrenheit and Celsius scales are used in the United States. The Celsius scale, part of the metric system, is sometimes called the Centigrade scale.

The line graph shows the relationship between Fahrenheit and Celsius scales.

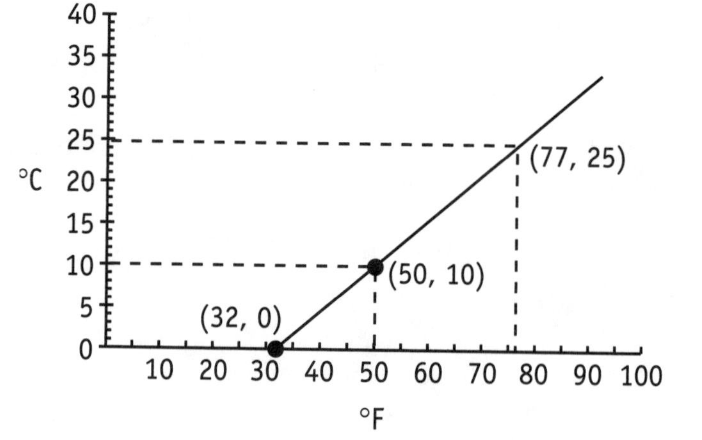

- Equivalent temperatures are plotted as points on a straight line.
- *Degrees Fahrenheit* (°F) are read on the horizontal axis.
- *Degrees Celsius* (°C) are read on the vertical axis.
- The coordinates of a point on the line give equivalent temperatures.

EXAMPLES 32°F = 0°C
50°F = 10°C

Refer to the graph and answer each question.

1. What does "°C" stand for?

2. What does "°F" stand for?

3. Find the approximate Celsius temperature for each Fahrenheit temperature. The first problem is done as an example.

   a. 32°F = *0°C* b. 60°F ≈ c. 77°F ≈ d. 90°F ≈

On the graph at the right, °C are read on the horizontal axis; °F are read on the vertical axis.

4. Complete the line graph. Plot several points of equivalent temperatures. Draw a straight line through the plotted points. Use the graph above for data points.

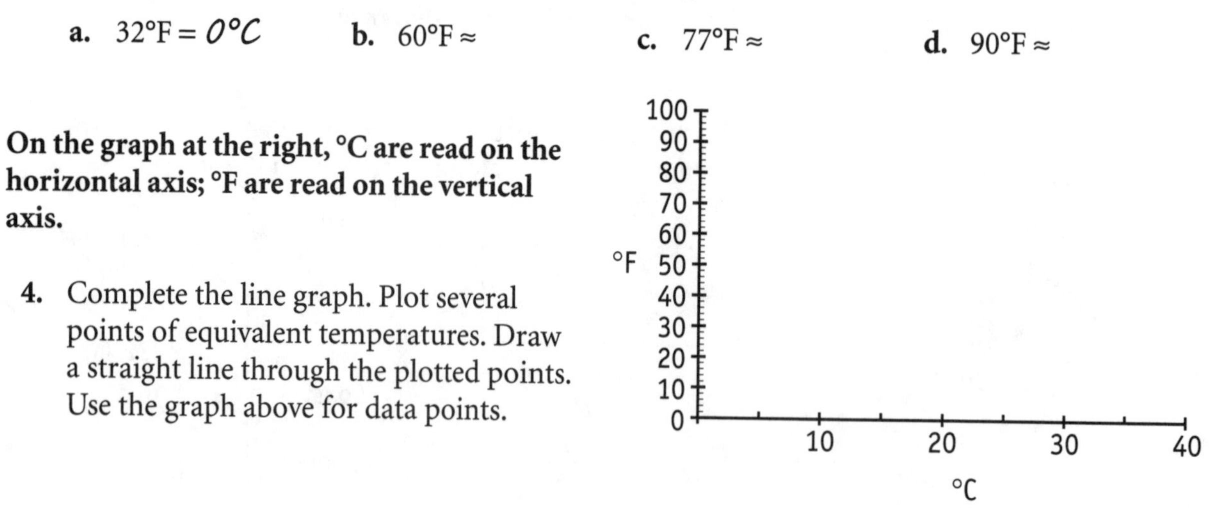

Celsius Temperature Formula

The Celsius temperature formula can be used to find the Celsius temperature when you know the Fahrenheit temperature.

Celsius temperature formula: $°C = \frac{5}{9}(°F - 32°)$

EXAMPLE Find the Celsius temperature when the Fahrenheit temperature is 98.6°F (average human body temperature).

$°C = \frac{5}{9}(°F - 32°)$

STEP 1 Substitute 98.6° for °F.

$°C = \frac{5}{9}(98.6° - 32°)$

STEP 2 Evaluate the expression within parentheses.
$98.6° - 32° = 66.6°$

$°C = \frac{5}{9}(66.6°)$

STEP 3 Multiply 66.6° by $\frac{5}{9}$.
$\frac{5}{9} \times 66.6° = \frac{333°}{9} = \mathbf{37°}$

$°C = \frac{333°}{9}$

$°C = \mathbf{37°}$

ANSWER: 37°C

Use the Celsius temperature formula to answer the following questions.

5. Carrie set her oven temperature at 350°F. What is the equivalent temperature in °C? Round your answer to the nearest degree.

6. The freezing point of water is 32°F. What is the freezing point of water when measured with a Celsius thermometer?

7. A comfortable room temperature is 70°F. What is the equivalent temperature given as a Celsius reading? Round your answer to the nearest degree.

Using a Calculator

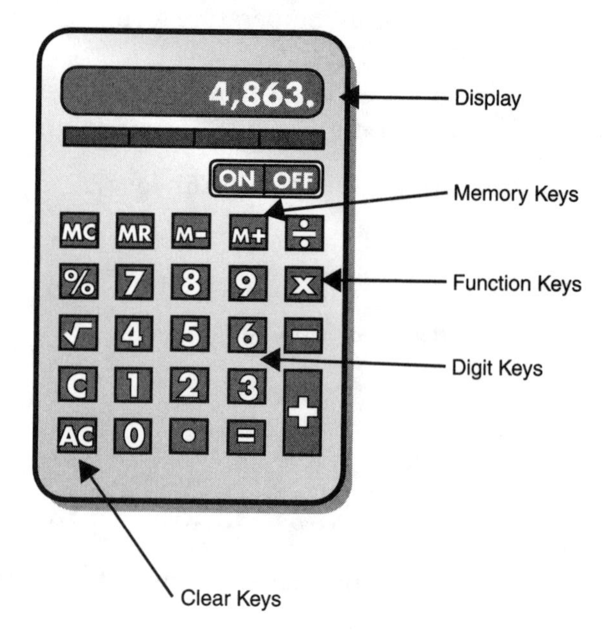

A calculator is a valuable math tool. You'll use it mainly to add, subtract, multiply, and divide quickly and accurately. The calculator pictured at the right is similar to one you've seen or may be using.

To enter a number, press one digit key at a time. On the display shown, the number 4,863 is entered. Notice the following features.

- A decimal point is displayed to the right of the ones digit.
- A comma separates groups of digits.
- The calculator does not have a comma [,] key or a dollar sign [$] key.
- Pressing the clear key [C] erases the display. You should press [C] each time you begin a new problem or when you've made a mistake.

EXAMPLE 1 Divide 408 by 12.

STEP 1 Press [C] to clear the display.

STEP 2 Press the digit keys, division key [÷], and equals key [=].

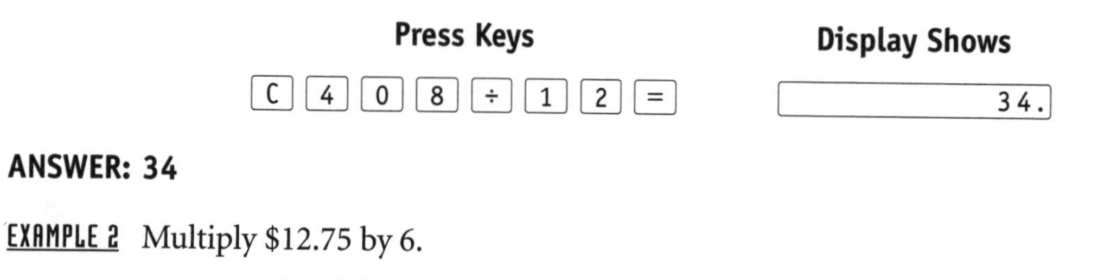

ANSWER: 34

EXAMPLE 2 Multiply $12.75 by 6.

STEP 1 Press [C] to clear the display.

STEP 2 Press the digit keys, decimal point key, times key [×], and equals key [=].

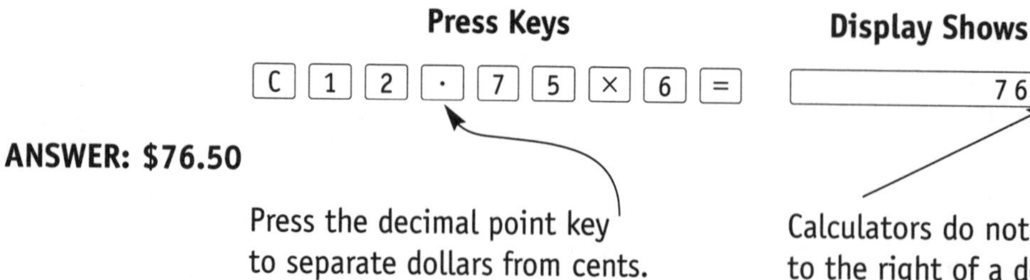

ANSWER: $76.50

Using Estimation and Mental Math

To **estimate** is to compute an approximate answer. Estimating is often done as **mental math**—math done in your head. One way to estimate that works well is to use rounded numbers that are easy to work with. You can estimate to

- discover about what an exact answer should be
- help pick a correct answer from among multiple choices—such as on multiple choice questions found on many tests
- quickly check an answer obtained when using a calculator—making sure you did not make a keying error

In *Pre-Algebra* there are questions for which you are asked to estimate an answer. On these questions, you do not need to calculate an exact answer. Here are some general guidelines for estimating using rounded numbers.

Whole Numbers

To estimate with whole numbers, replace each number with a rounded number.

EXAMPLE 1 Problem: 394×31

Estimate: $400 \times 30 =$ **12,000**

STEP 1 Round 394 to the nearest hundred.

STEP 2 Round 31 to the nearest ten.

Mixed Numbers

Round mixed numbers by replacing each mixed number with the nearest whole number.

EXAMPLE 2 Problem: $5\frac{1}{4} + 2\frac{7}{8} + 1\frac{1}{2}$

Estimate: $5 + 3 + 2 =$ **10**

STEP 1 Round $5\frac{1}{4}$ to 5.

STEP 2 Round $2\frac{7}{8}$ to 3.

STEP 3 Round $1\frac{1}{2}$ to 2.

Mixed Decimals

Round mixed decimals by replacing each mixed decimal with the nearest whole number.

EXAMPLE 3 Problem: 9.93×6.1

Estimate: $10 \times 6 =$ **60**

STEP 1 Round 9.93 to 10.

STEP 2 Round 6.1 to 6.

Formulas

To estimate with formulas, round whole numbers, mixed numbers, and mixed decimals as discussed above. Round π to 3 for a quick calculation.

EXAMPLE 4 Problem: $\pi \times 3\frac{7}{8} \times 3\frac{7}{8}$

Estimate: $3 \times 4 \times 4 =$ **48**

STEP 1 Round π to 3.

STEP 2 Round each $3\frac{7}{8}$ to 4.

Formulas

PERIMETER

Figure	Name	Formula	Meaning
	Rectangle	$P = 2l + 2w$	l = length w = width
	Square	$P = 4s$	s = side
	Triangle	$P = s_1 + s_2 + s_3$	s_1 = side 1 s_2 = side 2 s_3 = side 3
etc.	Polygon (n sides)	$P = s_1 + s_2 + \ldots s_n$	s_1 = side 1, and so on
	Circle (circumference)	$C = \pi d$ or $C = 2\pi r$	$\pi \approx 3.14$ or $\frac{22}{7}$ d = diameter r = radius

AREA

Figure	Name	Formula	Meaning
	Rectangle	$A = lw$	l = length w = width
	Square	$A = s^2$	s = side
	Parallelogram	$A = bh$	b = base h = height
	Triangle	$A = \frac{1}{2}bh$	b = base h = height
	Circle	$A = \pi r^2$	$\pi \approx 3.14$ or $\frac{22}{7}$ r = radius

VOLUME

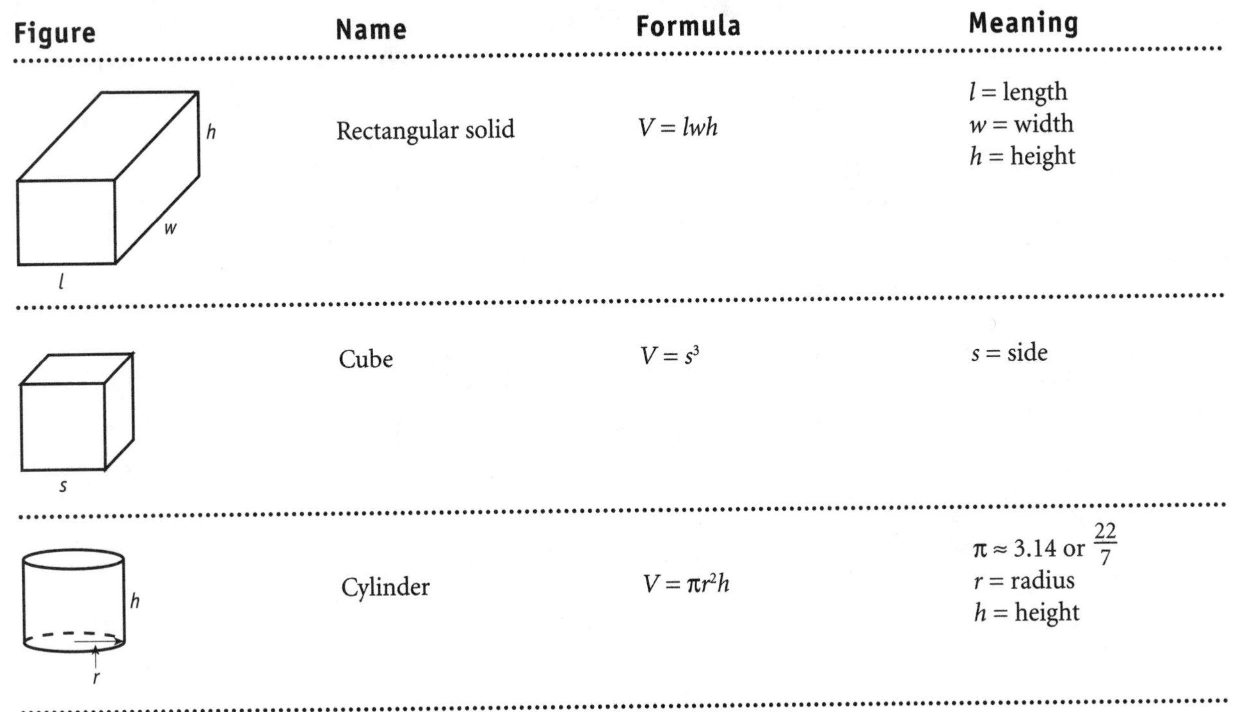

Figure	Name	Formula	Meaning
	Rectangular solid	$V = lwh$	l = length w = width h = height
	Cube	$V = s^3$	s = side
	Cylinder	$V = \pi r^2 h$	$\pi \approx 3.14$ or $\frac{22}{7}$ r = radius h = height

ANGLES

Name	Figure	Formula	Meaning
Sum of angles in a triangle		$\angle a + \angle b + \angle c = 180°$	a, b, and c are the three angles of any triangle.
Pythagorean theorem		$c^2 = a^2 + b^2$	c = hypotenuse a and b are the two shorter sides of a right triangle.

OTHER FORMULAS

Name	Formula	Meaning
Distance	$d = rt$	distance = rate × time
Average (Mean)	$(a + b + c + \ldots) \div n$	$a, b, c, \ldots$ are individual values in a set of values. n = the number of numbers in the set
Temperature	$°C = \frac{5}{9}(°F - 32°)$	°C means "degrees Celsius" °F means "degrees Fahrenheit"

Glossary

A

acute angle An angle measuring more than 0° but less than 90°

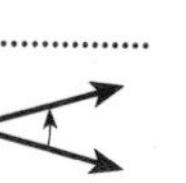
acute angle

adjacent angles Angles that share a side

algebraic expression Two or more numbers or variables combined by addition, subtraction, multiplication, or division

angle Figure formed when two rays are joined, forming a point called the vertex of the angle

angles

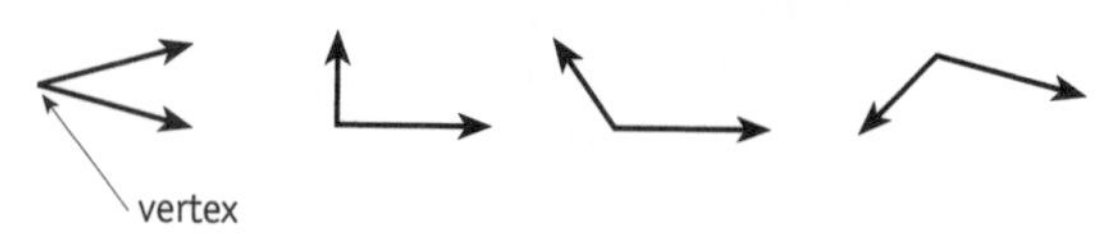

area (*A*) An amount of surface

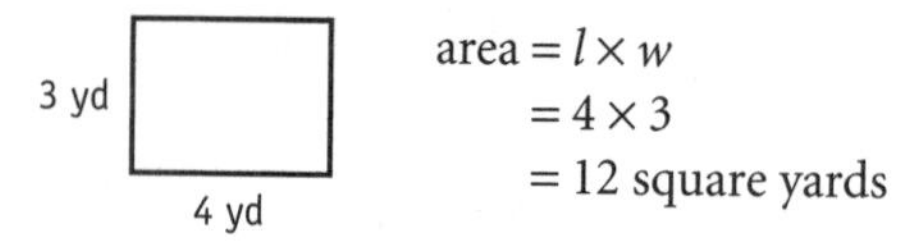

area $= l \times w$
$= 4 \times 3$
$= 12$ square yards

area unit Square units that are used to measure area. Common area units are square inches, square feet, square yards, and square meters.

area unit

average A typical or middle value of a set of values. The average (mean) is found by adding the numbers in the set and then dividing the sum by the number of values in the set. Another name for *average* is *mean.*

Stacey received three math scores: 83, 78, and 91.
Stacey's average score is 84.
$83 + 78 + 91 = 252$
$252 \div 3 = 84$

axis One of the perpendicular sides of a graph along which numbers, data values, or labels are written. Plural is *axes.*

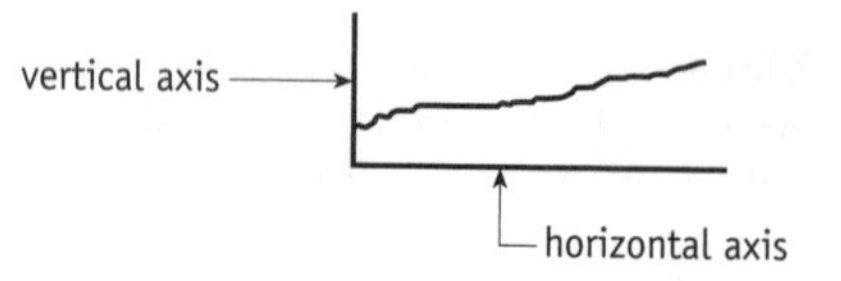

B

bar graph A graph that uses bars to display information. Data bars can be drawn vertically or horizontally.

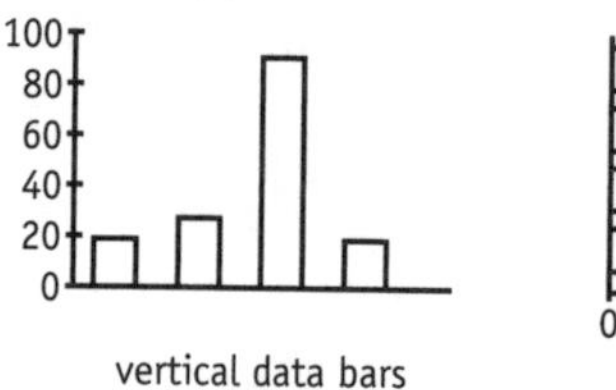

vertical data bars

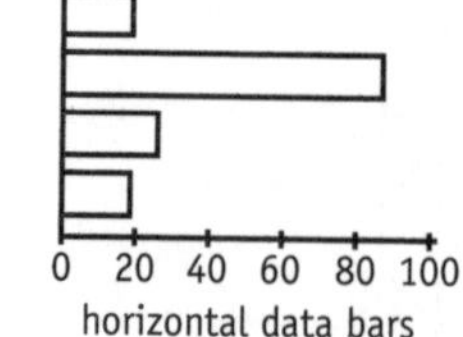

horizontal data bars

base The number being multiplied in a power. For example, in 4^3, 4 is the base.

base angles See *isosceles triangle.*

C

capacity The volume of liquid (or substance such as sugar) that a container can hold. Common capacity units are fluid ounce, quart, gallon, milliliter, and liter.

centimeter A meter unit of length equal to 0.01 meter
1 inch = 2.54 centimeters

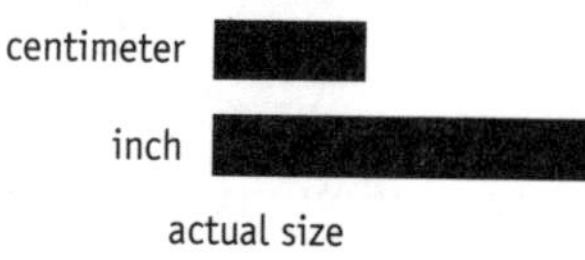

actual size

circle A plane (flat) figure, each point of which is an equal distance from the center

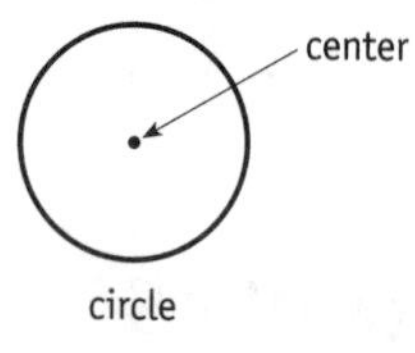

circle

circle graph A graph that uses a divided circle to show data. Each part of the circle is called a *part* or a *section.* The sections add up to a whole or to 100%. A circle graph is also called a *pie graph* or *pie chart.*

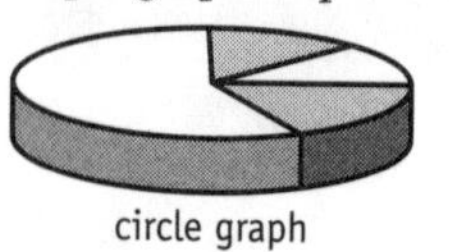
circle graph

circumference The distance around a circle. Circumference = $\pi \times$ diameter ($\pi \approx 3.14$)

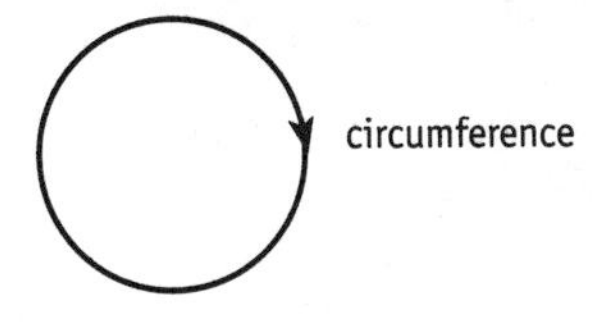

column A vertical listing that is read from top to bottom

comparison symbols Symbols used to compare one number with another. Examples: $<$ and $>$

Symbol	Meaning	Example	
$<$	is less than	$5 < 8$	5 is less than 8.
$>$	is greater than	$7 > -3$	7 is greater than −3.
$\leq$	is less than or equal to	$x \leq 9$	x is less than or equal to 9.
$\geq$	is greater than or equal to	$n \geq -3$	n is greater than or equal to −3.
$\neq$	is not equal to	$s \neq 6$	s is not equal to 6

complementary angles Angles whose sum is 90°

$\angle$A and $\angle$B are complementary angles.

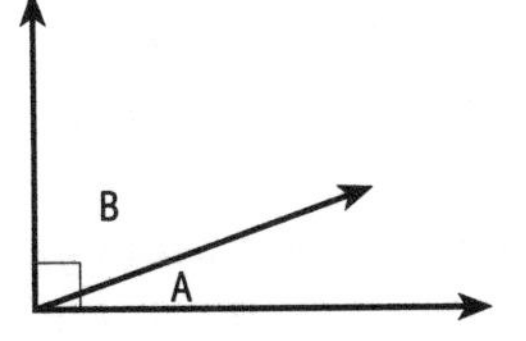

composite number A number that has more than two factors

congruent Having exactly the same size and shape

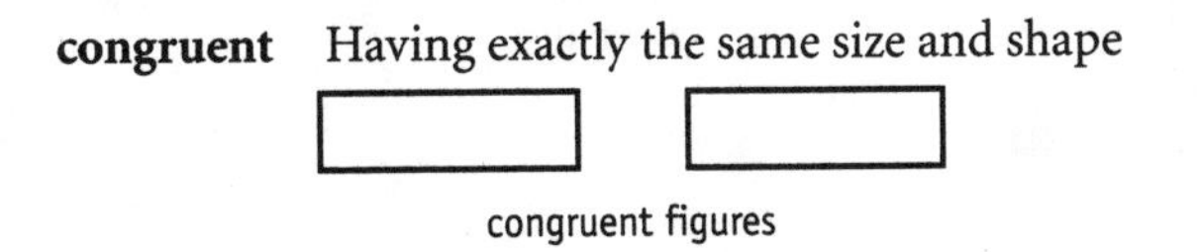

congruent figures

coordinate axis A number line used as either the horizontal or vertical axis of a graph. See *horizontal axis* and *vertical axis.*

corresponding angles Pairs of equal angles in similar triangles

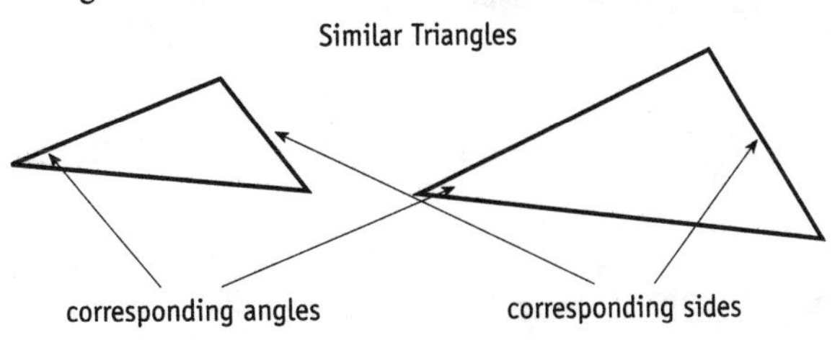

corresponding sides In similar triangles, sides that are across from corresponding angles. See *corresponding angles.*

cross products In a proportion, the product of the numerator of one ratio times the denominator of the other. In a true proportion, cross products are equal.

Proportion	Equal Cross Products
$\frac{2}{3} = \frac{6}{9}$	$2 \times 9 = 3 \times 6$
	$\mathbf{18 = 18}$

cube A 3-dimensional shape that contains six square faces. At each vertex, all sides meet at right angles.

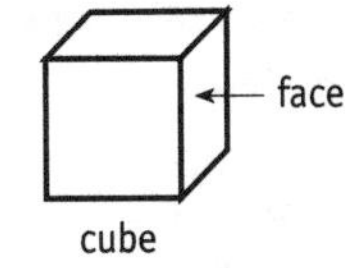

cylinder A 3-dimensional shape that has both a circular base and a circular top.

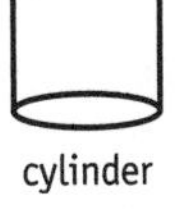

D

data A group of names, numbers, or other information that is related in some way

Number data: $2.50, $3.75, $6.40
Word data: beef, chicken, fish, pork

data analysis Organizing, interpreting, and using data. Also called *statistics.*

degrees The measure (size) of an angle. A circle contains 360 degrees (360°). One-fourth of a circle contains 90°.

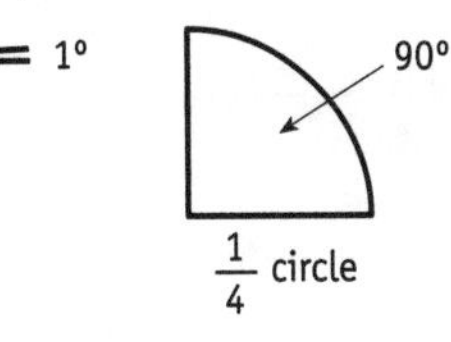

diagonal A line segment running between two non-consecutive vertices of a polygon. A diagonal divides a square or rectangle into two equal triangles.

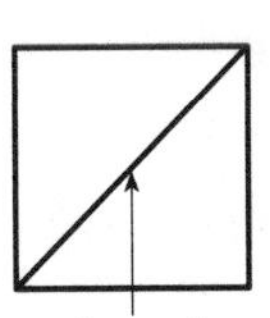

diameter A line segment, passing through the center, from one side of a circle to the other. The length of the diameter is the distance across the circle.

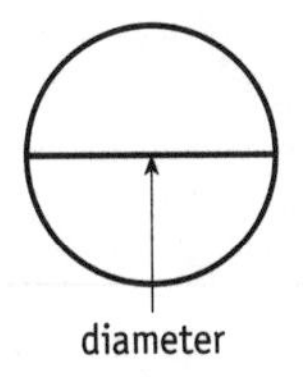

distance formula The formula $d = rt$ that relates distance, rate, and time. In words, distance equals rate times time. By rearranging the variables, you can write the rate formula $r = \frac{d}{t}$ or the time formula $t = \frac{r}{d}$.

E

equation A statement that two amounts are equal (have equal value or measure)

equilateral triangle A triangle with three equal sides and three equal angles, each measuring 60°. An equilateral triangle is a regular polygon.

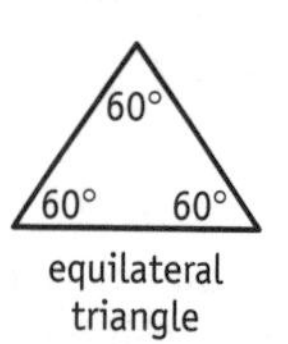

equilateral triangle

exponent A number that tells how many times the base (of a power) is written in the product. For example, in 5^2, 2 is the exponent.

estimate To find an approximate answer by calculating with rounded numbers

extrapolate To make a reasonable guess of a data value that lies outside a given set of values

Four given data values: 3, 6, 9, 12
Extrapolated fifth value: 15

F

factor A number that divides evenly into another number. Example: 1, 2, 4, and 8 are factors of 8.

favorable outcome In probability, a particular outcome (happening) that you are interested in

fluid ounce The smallest common capacity unit in the U.S. customary system. One cup is 8 fluid ounces; 1 quart is 32 fluid ounces.

foot A unit of length equal to 12 inches. There are 3 feet in 1 yard.

function keys On a calculator, keys that are used to perform mathematical operations. Function keys are used to add, subtract, multiply, and divide.

G

gallon A unit of capacity equal to 4 quarts. One gallon is 128 fluid ounces.

1 cubic foot ≈ $7\frac{1}{2}$ gallons

H

horizontal Running right and left

horizontal axis On a graph, the axis running left to right

horizontal axis

hypotenuse In a right triangle, the side opposite the right angle

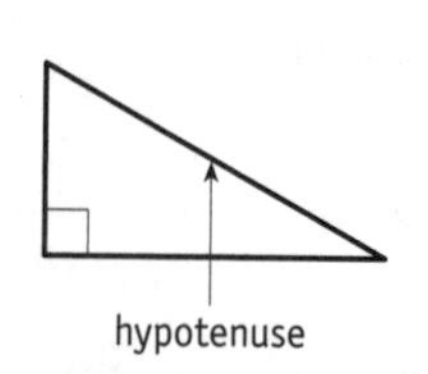

I

inequality A statement that compares two numbers. Examples: $n \geq 6$ states that n is greater than or equal to 6; $-2 < x \leq 3$ states that x is greater than −2 and less than or equal to 3.

intercept A point where a graphed line crosses a coordinate axis

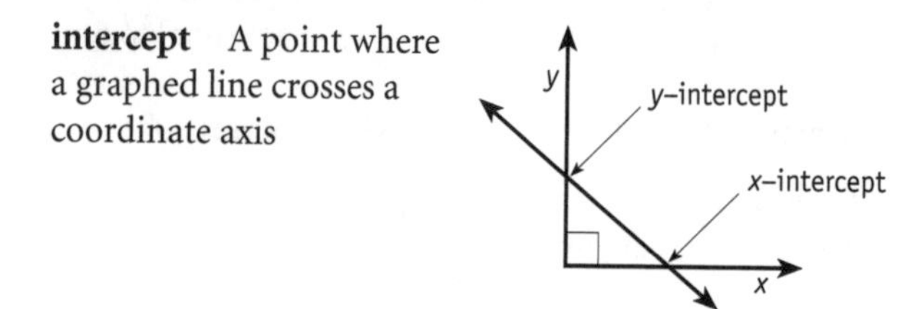

interpolate To estimate or guess a data that lies between two known values

isosceles triangle A triangle in which two sides have the same length. The two angles opposite the equal sides have the same measure. The two equal angles are called *base angles.*

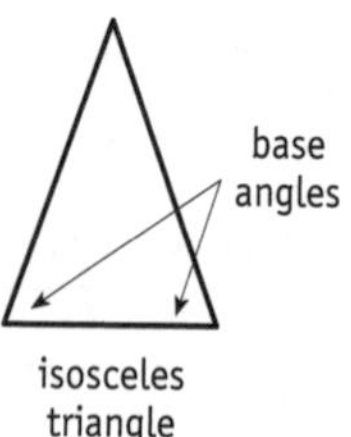

isosceles triangle

K

kilometer A metric unit of length equal to 1,000 meters. 1 meter ≈ 0.6 miles

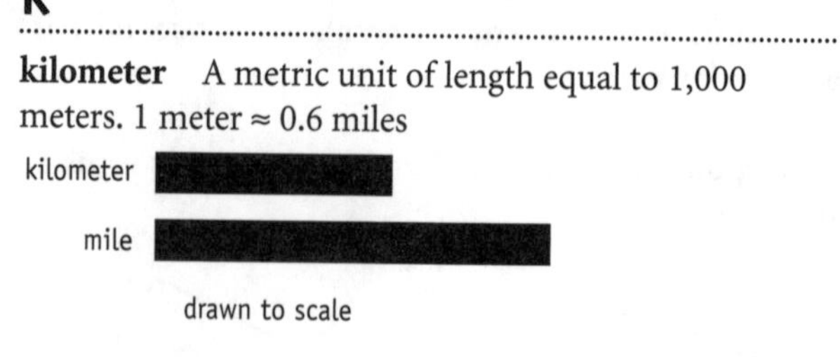

drawn to scale

L

line A path of points that extend in two opposite directions. A line has no endpoints.

line

line graph A graph that displays data as points along a graphed line

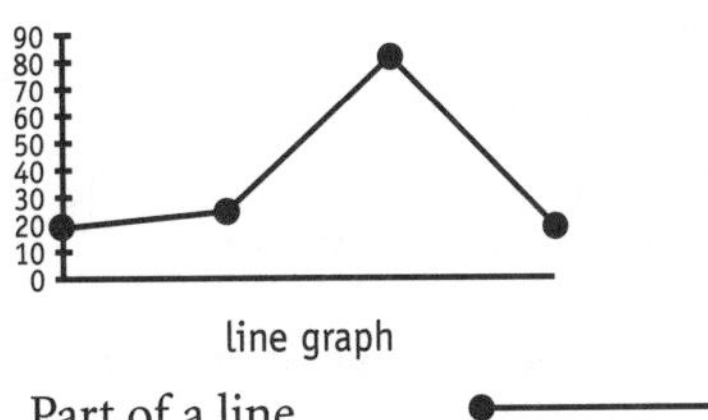

line graph

line segment Part of a line, having two endpoints

line segment

M

map directions Directions used on a map: north (N) at the top, south (S) at the bottom, west (W) at the left, and east (E) at the right.

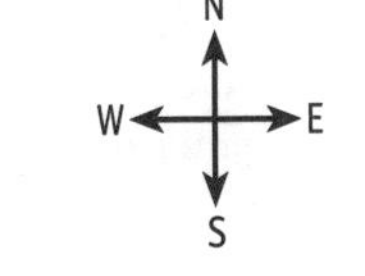

mean Equal to the sum of values in a set divided by the number of values. See *average.*

median The middle value of an uneven number of values in a set, or the mean of the two middle values of an even number of values

meter A metric unit of length equal to 100 cm. 1 meter ≈ 39.6 inches, a little longer than 1 yard

drawn to scale

metric system The measurement system used in most of the world. The United States is slowly adapting to metric measurement.

mode In a set, the value occurring most often

N

negative number A number less than zero. A negative number is written with a negative sign. Examples: –6, –2.5, –1

number line A line used to represent both positive and negative numbers. A number line can be written horizontally (left to right) or vertically (up and down).

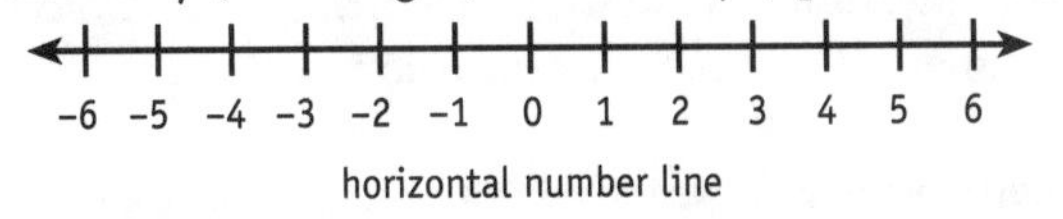

horizontal number line

numerical expression Two or more numbers combined by addition, subtraction, multiplication, or division

O

obtuse angle An angle greater than 90° but less than 180°

obtuse angle

P

parallel lines Lines that run side by side and do not cross

parallel lines

parallelogram A four-sided polygon with two pairs of parallel sides. Opposite sides are equal and opposite angles have equal measure.

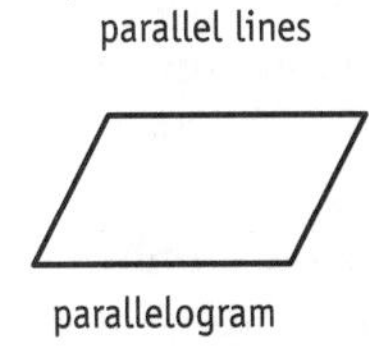
parallelogram

percent Part of 100. For example, 5 percent means 5 parts out of 100; 5¢ is 5% of $1.00.

perfect square A number whose square root is a whole number. Example: 49 is a perfect square. $\sqrt{49} = 7$

perimeter (*P*) The distance around a flat (plane) figure

perpendicular lines Lines that meet (or cross) at a right angle (90°)

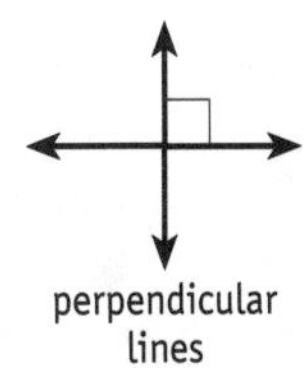
perpendicular lines

pi (π) The ratio of the circumference of a circle to its diameter. Pi is approximately 3.14 or $\frac{22}{7}$.

pictograph A graph that uses small pictures or symbols to represent data. Data lines may be displayed either horizontally or vertically.

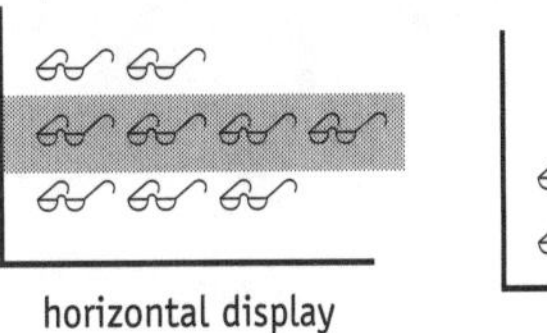
horizontal display

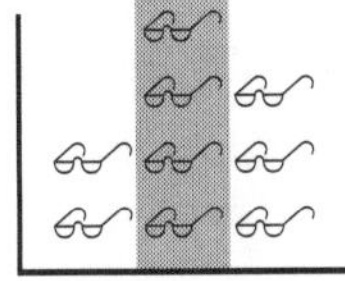
vertical display

pie chart or **graph** See *circle graph.*

plane figure A 2-dimensional (flat) figure. Examples: circles, squares, rectangles, and triangles

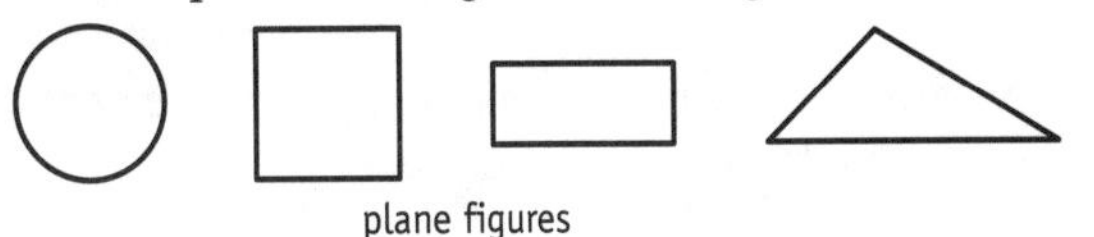
plane figures

polygon A plane (flat) figure formed by line segments that meet only at their endpoints

positive number A number greater than 0. Positive numbers may be written with a positive sign (+) or no sign at all.

power The product of a number multiplied by itself one or more times. Example: $3^2 = 3 \times 3$

prediction Regarding a graph, a guess about the value of an unknown data point

prime number Any number greater than 1 that has only two factors: itself and the number 1

prime-factorization form Writing a number as a product of prime factors

Examples: $10 = 2 \times 5$; $72 = 2 \times 2 \times 2 \times 3 \times 3 = 2^3 3^2$

probability The study of chance—the likelihood of an event happening

proportion Two equal ratios. A proportion can be written with colons or as equal fractions.

Written with colons: 2:3 = 6:9

Written as equal fractions: $\frac{2}{3} = \frac{6}{9}$

Pythagorean theorem A theorem that states: In a right triangle the square of the hypotenuse is equal to the sum of the squares of the two remaining sides.

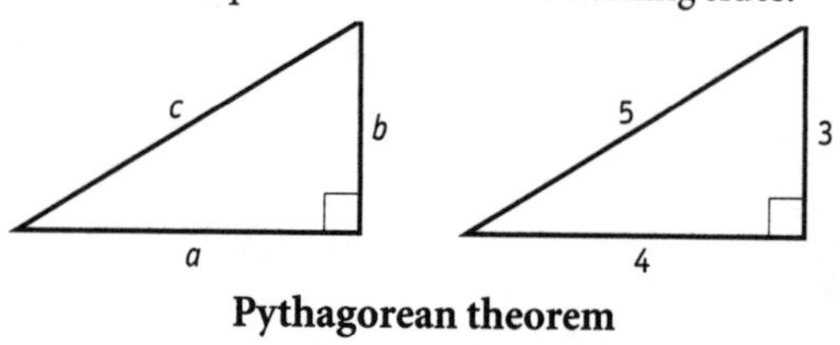

Pythagorean theorem

$c^2 = a^2 + b^2$

$5^2 = 4^2 + 3^2$
$25 = 16 + 9$

R

radius A line segment from the center of a circle to any point on the circumference of the circle

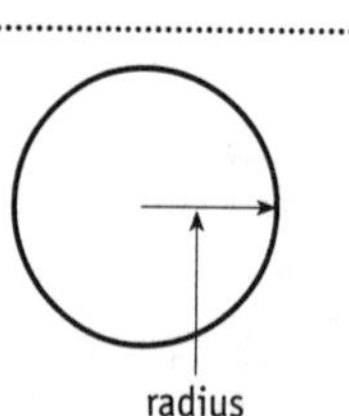

rate A ratio that compares two types of measurement. For example, a rate of 50 miles per hour

ratio A comparison of two numbers. For example, 5 to 4. A ratio can be written as a fraction or with a colon.

5 to 4 is written $\frac{5}{4}$ or 5:4

ray Part of a line. A ray has one endpoint. An angle is formed when two rays are joined at their endpoints.

endpoint

ray

rectangle A four-sided polygon with two pairs of parallel sides and four right angles

rectangle

rectangular solid (prism) A 3-dimensional figure in which each face is either a rectangle or a square. Opposite faces are congruent.

rectangular solid

reflex angle An angle that measures more than 180° but less than 360°

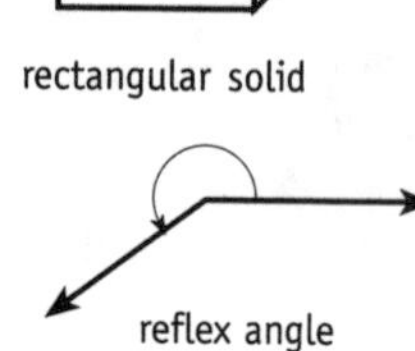

repeating decimal A decimal with a never-ending, repeating pattern of one or more digits. Also called a non-terminating decimal. Example: $0.54545454\ldots = 0.\overline{54}$

right angle An angle that measures exactly 90°. A right angle is often called a *corner angle.*

right angle

right triangle A triangle that contains a right angle

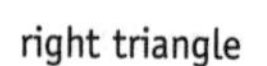

row A horizontal list, read from left to right

row of numbers: 14, 19, 26, 31
row of words: pennies, nickels, dimes, quarters

S

set A group of numbers, names, dates, and so on

set-up question A question in which answer choices are numerical or algebraic expressions

signed numbers Negative and positive numbers

similar triangles Triangles that have three equal angles. Similar triangles have the same shape and differ only in the lengths of their sides.

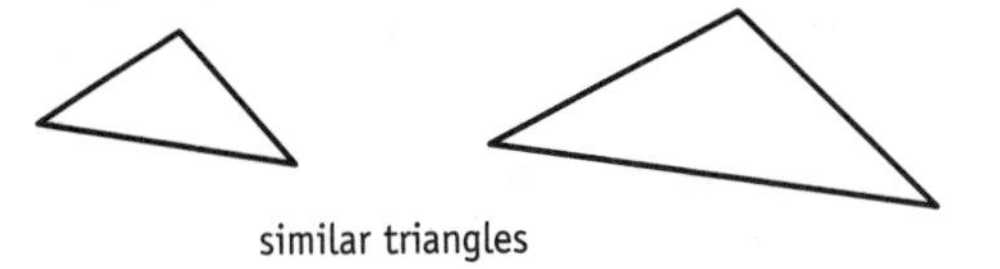
similar triangles

square A polygon with four equal sides, two pairs of parallel sides, and four right angles

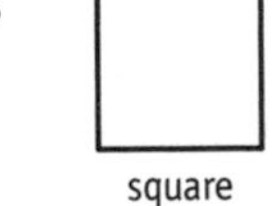
square

square of a number The product of a number multiplied by itself. Example: $8^2 = 8 \times 8 = 64$

square root ($\sqrt{\ }$) One of two equal factors of a number. Example: $\sqrt{9} = 3$

statistics See *data analysis.*

straight angle An angle that measures exactly 180°, having the shape of a straight line

straight angle

supplementary angles Angles that add up to 180°

C D

$\angle C$ and $\angle D$ are supplementary angles.

T

tally mark A slash that represents a count. Examples: || stands for 2; ~~||||~~ stands for 5.

tally sheet A record of tally marks

terminating decimal A decimal with a limited number of digits. Example: 0.625

three-dimensional figures Figures that take up space and have volume. Examples: cubes, rectangular solids, and cylinders

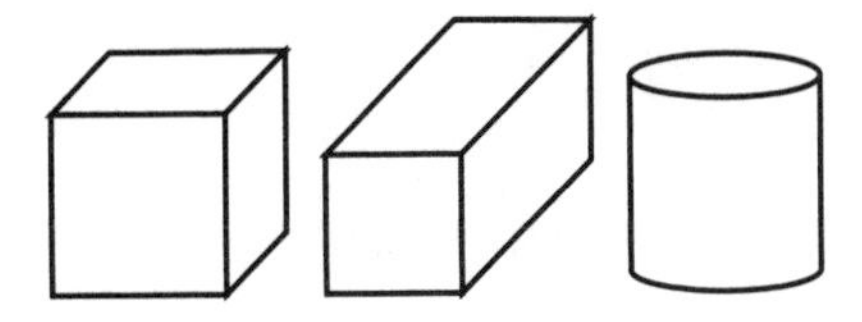

transversal A line that cuts across parallel lines, intersecting each of them

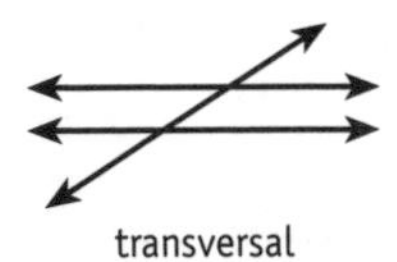
transversal

tree diagram A diagram, resembling a branching tree, that shows the number of possible combinations

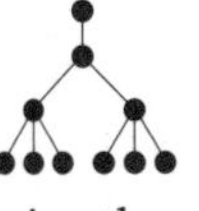

triangle A polygon having three sides and three angles

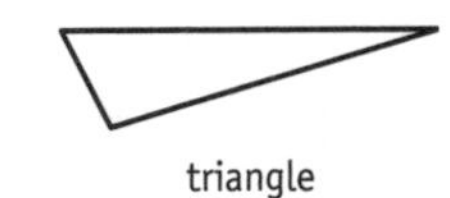
triangle

two-dimensional figure See *plane figure.*

V

vertex The point where the two sides of an angle meet. Plural is **vertices.**

vertical Running up and down or top to bottom

vertical angles Angles formed when two lines cross. Vertical angles lie across from each other and are equal.

E F

$\angle E$ and $\angle F$ are vertical angles: $\angle E = \angle F$

vertical axis On a graph, the axis running up and down

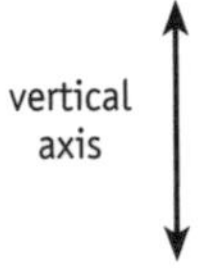
vertical axis

volume (*V*) The amount of space an object takes up. Volume is usually measured in cubic units.

volume unit Cubic units used to measure volume. Common volume units are cubic inches, cubic feet, cubic yards, cubic centimeters, and cubic meters.

Index

P

R

S

T

V